Public concern regarding environmental pollution and chemicals present in foods, consumer products, and the work place is at an all time high. Whilst there is widespread awareness, confusion still reigns, aggravated by conflicting reports concerning carcinogens in food and drinking water, or about chemicals present in medicines and household products that may cause birth defects. The effort to understand how these pollutants and chemical products may harm human health is led by scientists in the disciplines of toxicology, epidemiology and risk assessment.

The central purpose of this book is to describe how scientists come to understand the toxic properties of such chemicals and the health risks they may pose. Rather than attempting to expose governmental and corporate ignorance, negligence or corruption, this book explores the underlying scientific issues. It presents a practical and balanced clarification of the scientific basis for our concerns and uncertainties. It should serve to refocus the debate.

CALCULATED RISKS

Understanding the toxicity and human health risks
of chemicals in our environment

CALCULATED RISKS

Understanding the toxicity and human health risks of chemicals in our environment

Joseph V. Rodricks

Environ Corporation
Counsel in Health and Environmental Science

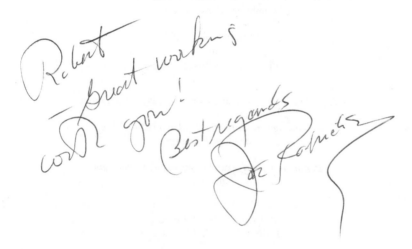

CAMBRIDGE
UNIVERSITY PRESS

Published by the Press Syndicate of the University of Cambridge
The Pitt Building, Trumpington Street, Cambridge CB2 1RP
40 West 20th Street, New York, NY 10011–4211, USA
10 Stamford Road, Oakleigh, Melbourne 3166, Australia

First published 1992
Reprinted 1993
First paperback edition 1994

Printed in Great Britain at the University Press, Cambridge

*A catalogue record of this book is available from
the British Library*

*Library of Congress cataloguing in publication data
available*

ISBN 0 521 41191 2 hardback
ISBN 0 521 42331 7 paperback

This book is dedicated to

Joseph A. Rodricks (1914–1986)
and
Rose M. Rodricks

Contents

Preface

Think how many carcinogens are household names: asbestos, cigarette smoke (a mixture of several thousand chemical compounds), DES, dioxin, saccharin, arsenic, PCBs, radon, EDB, Alar. Hundreds more of these substances, some very obscure, are known to the scientific and medical community, and many of these are scattered throughout the land at thousands of hazardous waste sites similar to Love Canal. People are exposed to these dreadful substances through the air they breathe, the water they drink and bathe in, and the foods they eat. Chemicals can also produce many other types of health damage, some very serious, such as birth defects and damage to our nervous and immune systems.

The chemical accident at Bhopal, India, in late 1984, is only the worst example of events that take place almost daily, on a smaller scale, throughout the world. Human beings are not the only potential victims of chemical toxicity – all of life on earth can be affected. Chemicals are ravaging human health and the environment, and conditions are worsening.

But wait. Let's remember that chemicals have virtually transformed the modern world in extraordinarily beneficial ways. During the past 100 years the chemical industry has offered up, and we have eagerly consumed, thousands of highly useful materials and products. Among these products are many that have had profoundly beneficial effects on human health – antibiotics and other remarkable medicinal agents to prevent and cure diseases, pesticides to protect crops, preservatives to protect the food supply, plastics, fibers, metals and hundreds of other materials that have enhanced the safety and pleasures of modern life.

Perhaps the misuse of certain chemicals has caused some small degree of harm, but on balance the huge benefits of modern chemical society clearly outweigh the exceedingly small risks these products may carry. Moreover, we have made and are continuing to make progress in controlling the risks of chemical technology.

Somewhere between these two views sits a somewhat befuddled scientific and medical community, attempting to sort the true from the false, and not quite sure how it should respond to the public while this sorting takes place. What science can now say with reasonably high certainty about the risks of chemical technology falls far short of the knowledge about those risks that our citizens are seeking. And, although substantial progress in scientific understanding has been made during the past three to four decades, it will probably be another several decades before the questions about chemical risks facing us today can be answered with the degree of certainty normally sought by scientists.

It is not at all surprising that confusion and controversy should arise when knowledge is absent or weak. When, as in the case of the risks of the products and byproducts of chemical technology, scientists know just enough to raise fearful suspicions, but do not always know enough to separate the true fears from the false, other social forces take command. Among the most important of these forces are the environmental laws that sometimes require regulatory authorities to act even before scientific understanding is firm. When the consequences of these actions cause economic harm, combat begins. Depending on which side of the battle one sits, fears about chemical risks are emphasized or downplayed. The form of the battle that will occur following what have become routine announcements about carcinogens in pancakes or apples, or nervous system poisons in drinking water or soft drinks, is now highly predictable. Except for a few brave (or foolish?) souls, the scientific community tends to remain relatively impassive in such circumstances, at most calling for 'more research'. Those scientists who are sufficiently intrepid to offer opinions tend to be scorned either as environmentalist quacks or industry hacks, who have departed from the traditional, scientifically acceptable standards of proof. Perhaps they have, but as we shall see, there is certainly an argument to be made on their behalf.

The question of whether and to what degree chemicals present in air, food, drinking water, medicinal agents, consumer products, and in the work place pose a threat to human health is obviously of enormous social and medical importance. This book is an attempt to answer this question with as much certainty as science can currently offer. It is in part a book of popular science – that is, it attempts to provide for the layman a view of the sciences of toxicology and chemical carcinogenesis (considered by some a branch of toxicology). It describes how the toxic properties of chemicals are identified, and how scientists make judgments about chemical risks. What is known with reasonable certainty is separated from the speculative; the large gray areas of science falling between these extremes are also sketched out. Toxicology, the science of poisons, is such a rich and fascinating subject, that it deserves more widespread recognition on purely intellectual grounds. Because it is now such an important tool in public health and regulatory decision-making, it is essential that its elements be widely understood.

The focus of this book is on the methods and principles of toxicology and risk assessment, and not on particular toxic agents or on the scientists who have built the discipline. To emphasize specific agents and scientists would have resulted in too great a departure from the book's second aim – to cast a little light upon the difficult interaction between science and the development of public health and regulatory policies. What is of interest here is not the administrative detail of policy implementation, which can be a rather unlively topic, but the principles that have come to govern the interaction of a highly uncertain scientific enterprise with the social demand for definitive actions regarding matters of public health.

The purpose of this book is, then, to describe and to clarify the scientific reasons for our present concerns about chemicals in the environment; the strengths and weaknesses of our scientific understanding; and the interplay between science and public policy. Unlike most other works related to these subjects, it is not an attempt to expose governmental and corporate ignorance, negligence or corruption. There is no end to literature on this subject, much of it presenting an incomplete or biased view of current scientific understanding of the effects of chemicals on human health and the environment. Perhaps a little clarification of the scientific bases for our concerns and the uncertainties that accompany them, and of the dilemmas facing decision-makers, will serve to refocus and advance the debate.

A word about organization of topics is in order. First, it is important to understand what we mean when we talk about 'chemicals'. Many people think the term refers only to generally noxious materials that are manufactured in industrial swamps, frequently for no good purpose. The existence of such an image impedes understanding of

toxicology and needs to be corrected. Moreover, because the molecular architecture of chemicals is a determinant of their behavior in
biological systems, it is important to create a little understanding of the
principles of chemical structure and behavior. For these reasons, we
begin with a brief review of some fundamentals of chemistry.

The two ultimate sources of chemicals – nature and industrial and
laboratory synthesis – are then briefly described. This review sets the
stage for a discussion of how human beings become exposed to
chemicals. The conditions of human exposure are a critical determinant of whether and how a chemical will produce injury or disease, so
the discussion of chemical sources and exposures naturally leads to the
major subject of the book – the science of toxicology.

The major subjects of the last third of this volume are risk assessment
– the process of determining the likelihood that chemical exposures
have or will produce toxicity – and risk control, or management, and
the associated topic of public perceptions of risk in relation to the
judgments of experts. It is particularly in these areas that the scientific
uncertainties become most visible and the public debate begins to heat
up. The final chapter contains some suggestions for improving the
current state-of-affairs and also sets out some new challenges. Risks to
human health are the subject of this book; risks to the rest of the living
and non-living environment are not covered. The absence of this topic
from the present volume has only to do with the author's interests and
knowledge and says nothing about its relative importance.

Much of the discussion of risk assessment turns on the activities of
regulatory agencies responsible for enforcing the two dozen or so
federal laws calling for restrictions of one sort or another on human
exposures to environmental chemicals. The Environmental Protection
Agency (EPA) has responsibilities for air and water pollutants, pesticides, hazardous wastes, and industrial chemicals not covered by other
statutes. The Food and Drug Administration (FDA), part of the
Department of Health and Human Services, manages risks from foods
and substances added thereto, drugs for both human and veterinary
uses (some of the latter can reach people through animal products such
as meat, milk and eggs), cosmetics, and constituents of medical devices.
The Occupational Safety and Health Administration (OSHA), a unit of
the Labor Department, handles chemical exposures in the workplace.
Consumer products not covered by other agencies fall to the Consumer
Product Safety Commission (CPSC). Other agencies with similar,
though somewhat narrower responsibilities, include the Food Safety
and Inspection Service of the Department of Agriculture (for meat,

poultry, and eggs) and the Department of Transportation. Although laws and regulatory programs vary, most countries have agencies with similar sets of responsibilities.

The use of the phrase 'regulatory risk assessments' throughout this book may seem odd, because risk assessment is a scientific activity and its conduct, it would seem, should be independent of where it is undertaken. But we shall see that a scientific consensus on the proper conduct of risk assessment does not exist, and regulatory agencies have had to adopt, as a matter of policy, certain assumptions that do not have universal acceptance in the scientific community. The agencies do this to allow them to operate in accordance with their legal mandates, and one of the purposes of this book is to create understanding of (but not necessarily to urge agreement with) these regulatory policies.

Before embarking on what is perhaps an overly systematic approach to our subject, we should attempt to develop a bird's-eye view of the entire landscape. We shall use a specific example – the case of a group of chemicals called aflatoxins – to illustrate the type of problem this book is designed to explore.

Prologue – turkeys, groundnuts, and cancer

In the fall of 1960 thousands of turkey poults and other animals started dying throughout southern England. Veterinarians were at first stymied about the cause of what they came to label 'Turkey X Disease', but because many thousands of birds eventually succumbed to the condition, a major investigation into its origins was undertaken. In 1961, a report from three scientists at London's Tropical Products Institute and a veterinarian at the Ministry of Agriculture's laboratory at Weybridge, entitled 'Toxicity associated with certain samples of groundnuts,' was published in the internationally-recognized scientific journal, *Nature*. Groundnuts, as everyone in America knows, are actually peanuts, and peanut meal is an important component of animal feed. It appeared that the turkeys had been poisoned by some agent present in the peanut meal component of their feed. The British investigators found that the poisonous agent was not a component of the peanuts themselves, but was found only in peanuts that had become contaminated with a certain mold.

It also became clear that the mold itself – identified by the mold experts (mycologists) as the fairly common species *Aspergillus flavus* – was not directly responsible for the poisoning. Turkey X Disease could be reproduced in the laboratory not only when birds were fed peanut meal contaminated with living mold, but also when fed the same meal after the mold had been killed.

Chemists have known for a long time that molds are immensely productive manufacturers of organic chemical agents. Perhaps the best known mold product is penicillin, but this is only one of thousands of such products that can be produced by molds. Why molds are so good at chemical synthesis is not entirely clear, but they surely can produce an array of molecules whose complexities are greatly admired by the organic chemist.

In fact Turkey X disease was by no means the first example of a mold-related poisoning. Both the veterinary and public health literature contain hundreds of references to animal and human poisonings associated with the consumption of feeds or foods that had molded, not only with *Aspergillus flavus*, but with many other species of mold as well. Perhaps the largest outbreaks of human poisonings produced by mold toxins occurred in areas of the Soviet Union just before and during the Second World War. Cereal grains left in the fields over the winter, for lack of sufficient labor to bring them in, became molded with certain varieties that grow especially well, and produce their toxic products, in the cold and under the snow. Consumption of molded cereals in the following springtime led to massive outbreaks of human poisonings characterized by hemorrhaging and other dreadful effects. The Soviet investigators dubbed the disease alimentary toxic aleukia (ATA). The mold chemicals, or *mycotoxins* ('myco' is from the Greek word for fungus, mykes), responsible for ATA are now known to fall into an extremely complex class of organic molecules called trichothecenes, although toxicologists are still at work trying to reconstruct the exact causes of this condition. Veterinary but probably not human poisonings with this class of mycotoxins still take place in several areas of the world.

Even older than ATA is ergotism. Ergot poisoning was widespread in Europe throughout the Middle Ages, and has occurred episodically on a smaller scale many times since. The most notable recent outbreak occurred in France in 1951. This gruesome intoxication is produced by chemical products of *Claviceps purpurea*, a purple-colored mold that grows especially well on rye, wheat, and other grains. Most of the ergot chemicals are in a class called alkaloids, one member of which is a simple derivative of the hallucinogenic agent, LSD, which of course came into popular use as a recreational drug during the 1960s. Ergot poisons produce a wide spectrum of horrible effects, including extremely painful convulsions, blindness, and gangrene. Parts of the body afflicted with gangrenous lesions blacken, shrink, dry up and may even fall off. The responsible mold is, unlike many others, fairly easy to spot, and normal care in the processing of grain into flour can eliminate the problem.

These and dozens more cases of mycotoxin poisonings were known to the investigators at the time they began delving into the causes of

Turkey X disease, so finding that a mold toxin was involved was no great surprise. But some new surprises were in store.

Investigations into the identity of the chemical agent responsible for Turkey X disease continued throughout the early 1960s at laboratories in several countries. At the Massachusetts Institute of Technology (MIT) a collaborative effort involving a group of toxicologists working under the direction of Gerald Wogan and a team of organic chemists headed by George Büchi had solved the mystery by 1965. The work of these scientists was a small masterpiece of the art of chemical and toxicological experimentation. After applying a long series of pain-stakingly careful extraction procedures to peanuts upon which the *Aspergillus flavus* mold had been allowed to grow, the research team isolated very small amounts of the substances that were responsible for the groundnut meal's poisonous properties. As is the custom among chemists, these substances were given a simple name that gave a clue to their source. Thus, from *Aspergillus flavus* toxin came the name aflatoxin.

Organic chemists are never satisfied with simply isolating and purifying such natural substances; their work is not complete until they identify the molecular structures of the substances they isolate. The case of aflatoxin presented a formidable challenge to the MIT team, because they were able to isolate only about 70 milligrams (mg) of purified aflatoxin with which to work (a mg is one-thousandth of a gram, and a gram is about 1/30th of an ounce). But the team overcame this problem through a masterful series of experimental studies, and in 1965 published details about the molecular structure of aflatoxin.

It turned out that aflatoxin was actually a mixture of four different but closely related chemicals. All possessed the same molecular back-bone of carbon, hydrogen, and oxygen atoms (which backbone was quite complex and not known to be present in any other natural or synthetic chemicals), but differed from one another in some minor details. Two of the aflatoxins emitted a blue fluorescence when they were irradiated with ultraviolet light, and so were named aflatoxins B_1 and B_2; the names aflatoxins G_1 and G_2 were assigned to the green-fluorescing compounds. The intense fluorescent properties of the aflatoxins would later prove invaluable aids to chemists interested in measuring the amounts of these substances present in various foods, because the intensity of the fluorescence was related to the amount of chemicals present.

While all this elegant investigation was underway, it became clear that the aflatoxins were not uncommon contaminants of certain foods.

A combination of the efforts of veterinarians investigating outbreaks of farm animal poisonings, survey work carried out by the Ministry of Agriculture in England, the U.S. Department of Agriculture (USDA) and the Food and Drug Administration (FDA), and the investigations of individual scientists in laboratories throughout the world, revealed during the 1960s and 1970s that aflatoxins can be found fairly regularly in peanuts and certain peanut products, corn grown in certain geographical areas, and even in some varieties of nuts. Cottonseed grown in regions of the Southwestern United States, but not in the Southeast, was discovered to be susceptible. While peanut, corn and cottonseed oils processed from contaminated products did not seem to carry the aflatoxins, these compounds did remain behind in the so-called 'meals' made from these products. These meals are fed to poultry and livestock and, if they contain sufficiently high levels of aflatoxins, the chemical agents can be found in the derived food products – meat, eggs, and especially milk. The frequency of occurrence of the aflatoxins and the amounts found vary greatly from one geographical area to another, and seem to depend upon climate and agricultural and food storage practices.

While this work was underway, toxicologists were busy in several laboratories in the United States and Europe attempting to acquire a complete profile of aflatoxins' poisonous properties. These substances did seem to be responsible for several outbreaks of liver poisoning, sometimes resulting in death, in farm animals, but there was no evidence that aflatoxins reaching humans through various food products were causing similar harm. The most likely reason for this lack of evidence was the fact that the amounts of aflatoxins reaching humans through foods simply did not match the relatively large amounts that may contaminate animal feeds. Of course, if aflatoxins were indeed causing liver disease in people, it would be extremely difficult to find this out unless, as in the case of ATA or ergotism, the signs and symptoms were highly unusual and occurring relatively soon after exposure.

In experimental studies in laboratory settings, aflatoxins proved not only to be potent liver poisons, but also – and this was the great surprise – capable of producing malignant tumors, sometimes in great abundance, in rats, ferrets, guinea pigs, mice, monkeys, sheep, ducks, and rainbow trout (trout are exquisitely sensitive to aflatoxin-induced carcinogenicity). Several studies from areas of the world in which human liver cancer rates are unusually high turned up evidence suggesting, but not clearly establishing, a role for the aflatoxins.

Aflatoxin's cancer-producing properties were uncovered and reported in the scientific literature during the period 1961–1976, the same period during which these substances were discovered to be low level but not infrequent contaminants of certain human foods.

What was to be done? Were the aflatoxins a real threat to the public health? How many cases of cancer could be attributed to them? Why was there no clear evidence that aflatoxins could produce cancers in exposed humans? How should we take into account the fact that the amounts of aflatoxins people might ingest through contaminated foods were typically very much less than the amounts that could be demonstrated experimentally to poison the livers of rodents and to increase the rate of occurrence of malignancies in these several species? And if aflatoxins were indeed a public health menace, what steps should be taken to control or eliminate human exposure to them? Indeed, because aflatoxins occurred naturally, was it possible to control them at all?

These and other questions were much in the air during the decade from 1965 to 1975 at the Food and Drug Administration – the public health and regulatory agency responsible for enforcing federal food laws and ensuring the safety of the food supply. FDA scientists and policy-makers consulted aflatoxin experts in the scientific community, food technologists in affected industries, particularly those producing peanut, corn, and dairy products, and experts in agricultural practices. The agency decided that limits needed to be placed on the aflatoxin content of foods. In the 1960s, FDA declared that peanut products containing aflatoxins in excess of 30 parts aflatoxin per billion parts of food (ppb) would be considered unfit for human consumption; a few years later the agency lowered the acceptable limit to 20 ppb. This ppb unit refers to the weight of aflatoxin divided by the weight of food; for one kilogram of peanut butter (about 2.2 lbs.), the 20 ppb limit restricts the aflatoxin content to 20 micrograms (one microgram is one-millionth of one gram – more will be said about these units later).

FDA's decision was based on the conclusion that no completely safe level of human intake could be established for a cancer-causing chemical. This position led, in turn, to the position that if analytical chemists could be sure aflatoxins were present in a food, then the food could not be consumed without threatening human health. The question, then, was the smallest amount of aflatoxin that analytical chemists could reliably detect: by 1968 this amount – or, more accurately, concentration – was 30 ppb, and because of improvements in analytical technology, the detection limit later dropped to 20 ppb. The analytical chemist dictated the FDA's position on acceptable aflatoxin limits.

It turned out that meeting a 20 ppb limit was not excessively burdensome on major manufacturers of peanut butter; aflatoxin tended to concentrate in discolored or otherwise irregular peanuts, which, fortunately, could be picked up and rejected by modern electronic sorting machines. Manufacturers did, however, have to institute substantial additional quality control procedures to meet FDA limits, and many smaller manufacturers had trouble meeting a 20 ppb limit. An extensive USDA program of sampling and analysis of raw peanuts, which continues to this day, was also put into place as the first line of attack on the problem. FDA's position on acceptable aflatoxin limits also took into account the capabilities of growers and processors to prevent or remove aflatoxin contamination.

Did this FDA position make any scientific sense? It implied that if aflatoxin could be detected by reliable analysis, it was too risky to be consumed by humans, but that if the aflatoxin happened to be present below the minimum detectable concentration it was acceptable. (Analytical chemists can never declare that a chemical is *not* present. The best that can be done is to show that it is not present above some level – 20 ppb in the case of aflatoxins, and other, widely varying, levels in the case of other chemicals in the environment.) To be fair to the FDA, perhaps the word 'acceptable' should be withdrawn; the agency's position was not so much that all concentrations of aflatoxin up to 20 ppb were acceptable, but that nothing much could be done about them because the chemists could not determine whether they were truly present in a given lot of food until the concentration exceeded 20 ppb.

Was FDA's position scientifically defensible? Let's offer two responses that might reflect the range of possible scientific opinion:

(1) Yes. FDA clearly did the right thing, and perhaps did not go far enough. Aflatoxins are surely potent cancer-causing agents in animals. We don't have significant human data, but this is very hard to get and we shouldn't wait for it before we institute controls. We know from much study that animal testing gives a reliable indication of human risk. We also know that cancer-causing chemicals are a special breed of toxicants – they can threaten health at any level of intake. We should therefore eliminate human exposure to such agents whenever we can, and, at the least, reduce exposure to the lowest possible level whenever we're not sure how to eliminate it.

or (2) No. The FDA went too far. Aflatoxins can indeed cause liver toxicity in
animals and are also carcinogenic. But they produce these adverse
effects only at levels far above the limit FDA set. We should ensure
some safety margin to protect humans, but 20 ppb is unnecessarily low
and the policy that there is no safe level is not supported by scientific
studies. Indeed, it's not even certain that aflatoxins represent a cancer
risk to humans because animal testing is not known to be a reliable
predictor of human risk. Moreover, the carcinogenic potency of
aflatoxins varies greatly even among the several animal species in
which they have been tested. Human evidence that aflatoxins cause
cancer is unsubstantiated. There's no sound scientific basis for FDA's
position.

The whole matter of protective limits for aflatoxin became more
complex in the early- to mid-1970s when it became clear that analytic
chemists could do far better than a 20 ppb detection limit. In several
laboratories, aflatoxins could easily be detected as low as 5 ppb, and in
some laboratories 1 ppb became almost routine. If FDA was to follow a
consistent policy, the agency would have had to call for these lower
limits. But it did no such thing. It had become obvious to FDA by the
mid-1970s that a large fraction of the peanut butter produced by even
the most technically advanced manufacturers would fail to meet a 1
ppb limit, and it was also apparent that other foods – corn meal and
certain other corn products, some varieties of nuts (especially Brazils
and pistachios) – would also fail the 1 ppb test pretty frequently. The
economic impact of a 20 ppb policy was not great. The impact of a 1
ppb limit could be very large for these industries. Did it still make
scientific sense to pursue an 'analytical detection limit' goal, at any
cost? Was the scientific evidence about cancer risks at very low intakes
that certain?

Here we come to the heart of the problem we shall explore in this
book: just how certain is our science on matters such as this? And how
should public health officials deal with the uncertainties? We shall be
exploring the two responses to FDA's position that were set out earlier
and learn what we can about their relative scientific merits, not
specifically in connection with the aflatoxin problem, but in a more
general sense. We shall also be illustrating how regulators react to these
various scientific responses, and others as well, using some examples
where the economic stakes are very high. One would like to believe that
the size of the economic stakes would not influence scientific thinking,
but it surely influences scientists and policy-makers when they deal
with scientific uncertainties.

In the meantime keep in mind that, although considerable progress has been made in reducing aflatoxin exposures, these mold products are still in some foods, and you have probably ingested a few nanograms (billionths of a gram!) recently. You'd surely like to know how much your health is threatened by these unusual compounds.

Acknowledgements

I was wise enough to have sought the assistance of several colleagues more expert than I on many of the topics I have covered, and fortunate enough to have received it. Several large and many small errors have been avoided because of careful review of the manuscript by Duncan Turnbull, Karen Wilcock, Benjamin Jackson and Larisa Rudenko. My indebtness to these four excellent scientists is large. Remaining errors are mine alone. Ema Yapobi-Attie labored unstintingly at the word processor and was of immense help to me, as was my secretary, Darlene Weathers, who never failed to assist my efforts. My wife Carol and daughter Elizabeth deserve special thanks both for enduring the sometimes unpleasant moods this effort provoked and for providing the support necessary for me to complete it.

1

Chemicals

Certain aspects of chemical science should be grasped before entering the domains of toxicology, risk assessment, and risk management. We need not dip into fundamental concepts and the lesson will be brief and of very limited scope. Many of the definitions and principles to be discussed are (or should be) well known by those who have passed through a high school chemistry course, even with relatively poor success, but we need to be reminded of them and keep them in mind, as matters of chemicals in the environment come under discussion later in the book.

It is perhaps too obvious to point out that everything we can see, touch, smell, and taste is a chemical or, more likely, a mixture of many different chemicals. In addition, there are many chemical substances in the environment that cannot be detected with the senses, but only indirectly, by the sophisticated instruments scientists have devised to look for them. The number of different chemicals in and on the earth is unknown, but is surely in the many millions. During the past 125 years scientists have been successful in creating hundreds of thousands of compounds that do not occur in nature, and they continue to add to the earth's chemical stores, although most of these synthesized chemicals never leave the research scientists' laboratories.

For both historical and scientific reasons chemists divide up the universe of chemicals into inorganic compounds and organic compounds. The original basis for classifying chemicals as 'organic' was the hypothesis, known since the mid-nineteenth century to be false, that organic chemicals could be produced only by living organisms. Modern scientists classify chemicals as 'organic' if they contain the element carbon.[1] Carbon has the remarkable and almost unique property that atoms of it can combine with each other in many different ways, and, together with a few other elements – hydrogen, oxygen, nitrogen, sulfur, chlorine, bromine, fluorine and a few more – can create a huge number of different molecular arrangements. Each such arrangement creates, of course, a unique chemical. Several million distinct organic chemicals are known to chemists, and there are many more that will be found to occur naturally or that will be created by laboratory synthesis. All of life – at least life on earth – depends on carbon compounds, and probably could not have evolved if carbon did not have its unique and extraordinary bonding properties, although chemists have verified many thousands of times over that the creation of organic chemicals does not depend on the presence of a living organism.[2]

Everything else is called inorganic. There are 90 elements in addition to carbon in nature (and several more that have been created in laboratories), and the various arrangements and combinations of these elements, some occurring naturally and others resulting from chemical synthesis, make up the remaining molecules in our world and universe. Because these elements do not have the special properties of carbon, the number of different possible combinations of them is smaller than can occur with carbon. The total number of known inorganic compounds is only a fraction of the total number of organic chemicals.

What is meant when a chemical is said to be 'known'? Typically this means that chemists have somehow isolated the substance from its source, have taken it to a relatively high state of purity (by separating it from chemicals that occur with it), have measured or evaluated its physical properties – the temperatures at which it melts, boils, and degrades, the types of solvents in which it dissolves, and so on – and have established its molecular architecture. This last act – determination of chemical architecture, or structure – typically presents the greatest scientific challenge. Understanding chemical structure – the number, type, and arrangement of atoms in a molecule – is important because structure determines how the compound undergoes change to

---

[1] There are a few compounds of carbon that chemists still consider inorganic: these are typically simple molecules such as carbon monoxide (CO) and carbon dioxide (CO_2) and the mineral limestone, which is calcium carbonate ($CaCO_3$).

[2] Of course the chemists creating organic compounds are living (most of them anyway) but the compounds are created in laboratory flasks without benefit of living organisms. Such synthesis is clearly different than, for example, the production of colors by flowers and aflatoxins by molds.

other compounds in chemical reactions, and also how it interacts with biological systems, sometimes to produce beneficial effects (nutrients and medicinal agents) and sometimes to produce harmful effects.

Chemists represent the structures of chemicals using letter symbols to represent atoms (C for Carbon, H for Hydrogen, O for Oxygen), and lines to indicate the chemical bonds that link atoms together (each bond is actually a pair of interacting electrons). The simplest of all organic compounds, the naturally occurring gas methane (marsh gas), has the structure represented below, in two dimensions.

$$
\begin{array}{c}
\text{H} \\
| \\
\text{H}-\text{C}-\text{H} \\
| \\
\text{H}
\end{array}
$$

Methane

Molecules of methane are actually three-dimensional, with the carbon atom at the center of a tetrahedron and a hydrogen atom at each of the four corners. When we refer to the chemical methane, we refer to collections of huge numbers of these specific molecules. An ounce of methane contains about 12×10^{23} (12 followed by 23 zeroes) of these molecules.

An interesting, important, and common phenomenon in organic chemistry is that of structural isomerism. Consider a molecule having two carbon atoms, four hydrogen atoms, and two chlorine atoms ($C_2H_4Cl_2$). These atoms are capable of binding to each other in two different ways as shown.

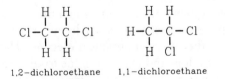

1,2–dichloroethane   1,1–dichloroethane

More ways are not possible because of limitations on the number of bonds each type of atom can carry (carbon has a limit of four, hydrogen and chlorine have a limit of one). But the important lesson here is that the two molecules shown (1,1-dichloroethane and 1,2-dichloroethane) are different chemicals. They have identical numbers of C, H, and Cl atoms ('isomers' means 'same weight') but different chemical structures.

These two chemicals have different physical and chemical properties, and even produce different forms of toxicity at different levels of exposure. The way chemicals interact with biological systems to produce damage depends greatly upon details of molecular structure.

The structures of inorganic compounds are represented by the same types of conventions as shown for the three organic compounds depicted above, but there are some important differences in the nature of the chemical bond that links the atoms. These differences need not concern us here, but will have to be mentioned at later points in the discussion. Toxicologists refer to many of the environmentally important inorganic chemicals simply according to the name of the particular portion of the compound that produces health damage. They refer, for example, to the toxicity of lead, mercury, or cadmium, without reference to the fact that these metals are actually components of certain compounds that contain other elements as well. Sometimes the particular form of the lead or mercury is important toxicologically, but often it is not. Toxicologists tend to simplify the chemistry when dealing with metals such as these.

Naturally occurring chemicals

Living organisms contain or produce organic chemicals, by the millions. The most abundant organic chemical on earth is cellulose, a giant molecule containing thousands of atoms of C, H, and O. Cellulose is produced by all plants and is the essential structural component of them. Chemically, cellulose is a carbohydrate (one that is not digestible by humans) which, together with proteins, fats, and nucleic acids are the primary components of life. But as mentioned in the Prologue in connection with the chemistry of molds, living organisms also produce huge numbers of other types of organic molecules. The colors of plants and animals, their odors and tastes, are due to the presence of organic chemicals. The numbers and structural varieties of naturally-occurring organic chemicals are staggering.

Other important natural sources of organic chemicals are the so-called fossil fuels – natural gas, petroleum, and coal – all deposited in the earth from the decay of plant and animal remains, and containing thousands of degradation products. Most of these are simple compounds containing only carbon and hydrogen (technically and even reasonably known as hydrocarbons). Natural gas is relatively simple in composition and is mostly made up of gases such as methane (marsh gas, already described above). The organic chemical industry depends upon these and just a few other natural products for everything it manufactures; the fraction of the fossil fuels not used directly for energy generation is used as 'feedstock' for the chemical industry. There are, or course, inorganic chemicals present in living organisms, many essential to life – the minerals. But the principal natural source of inorganic chemicals is the non-living part of the earth that humans have learned how to mine.

How people become exposed to chemicals will be discussed in the next chapter, but it is useful to note here that the greatest sources of chemicals to which we are regularly and directly exposed are the natural components of the plants and animals we consume as foods. In terms of both numbers and structural variations, no other chemical source matches food. We have no firm estimate of the number of such chemicals we are exposed to through food, but it is surely immense. A cup of coffee contains, for example, nearly 200 different organic chemicals – natural components of the coffee bean that are extracted into water. Some impart color, some taste, some aroma, others none of the above. The simple potato has about 100 different natural components (some quite toxic, as shall be seen), and to make matters more interesting and confusing, some of the chemicals found in the potato and the amounts present vary among different varieties and even different conditions of cultivation and storage! The issue of naturally-occurring food constituents will come up several times in this book.

Synthetic chemicals

The decade of the 1850s is noted by historians of science as significant because it saw the publication of Darwin's *Origin of Species* (1859), a work that has had profound influences on contemporary society. But other scientific events occurred at about the same time, which, I would argue if I were a historian, had nearly equal significance for our time. They allowed the development of organic chemical science and so greatly increased our understanding of chemical behavior that they spawned the age of chemical synthesis. Chemical synthesis is the science (one might say art) of building chemicals of specified structure from simpler and readily available chemicals, usually petroleum or coal products and other natural chemicals. Sometimes chemists engage in synthesis for rather obscure purposes, related to gaining some understanding of fundamental chemical principals. More often, they are interested in creating molecules that possess useful properties.

Chemists and their historical predecessors have for dozens of centuries been manipulating natural products to make useful materials, but for most of this time they had little understanding of what they were doing. It wasn't until the Structural Theory of Organic Chemistry began to solidify during the third quarter of the 19th century that chemists could possibly understand the molecular changes underlying such ancient arts as fermentation, dyeing, and soap making. The art of chemical purification had been well developed by the mid-19th century, but again chemists couldn't say much about the properties of the substances they had purified (mostly acids, alcohols, and aromatic chemicals from plants and animals) until structural theory came along. But once chemists grasped structural theory it became possible to manipulate chemicals in a systematic way so that certain molecular arrangements could be transformed in predictable ways to other, desired arrangements. The chemists who developed the structural theory of organic compounds and those who applied it to the synthesis of substances that were not products of nature, but totally new to the world, were instigators of a mammoth industrial revolution, one that has given us an extraordinary variety of beneficial materials.

Through the efforts of many chemical pioneers, mostly European, organic chemical science began to take on its contemporary shape during the first half of the nineteenth century. It was not until 1858, however, that Friedrich August Kekulé von Stradonitz, a student of architecture who had become captivated by chemical science and who held a position at the University of Heidelberg, and Archibald Scott Couper, a Scotsman then at the Sorbonne, independently introduced the so-called 'rules of valency' applicable to compounds of carbon. This work unified thinking about the structural characteristics of organic chemicals because it allowed chemists for the first time to explain hundreds of early observations on the chemical behavior of organic chemicals, and, as noted, structural theory also set the stage for the rapid development of the chemical industry. It is a distortion to say that Kekulé and Couper single-handedly formulated the structural theory of organic chemistry – they built upon and synthesized the earlier and quite extraordinary work of giants such as Edward Frankland (1825–99) at Manchester, Justus von Liebig (1803–73) at Geissen, Joens J.F. von Berzelius (1779–1848) at Stockholm, Friederick Wohler (1800–82) at Göttingen, Marcellin Bertholet (1827–1907) at

Paris, and several others as well. The publications by Kekulé and Couper in 1858 nevertheless clearly unleashed powerful new forces in chemical science and their societal repercussions have been profound, in practical ways perhaps more so than those produced by the Darwinian revolution.[3] Most history books do not seem to capture the work of these great scientists, nor have historians extensively explored their legacy for the modern world. Part of that legacy – the possible adverse public health consequences of their work – is the topic of this book.

The development of the chemical industry did not, of course, spring wholly from the work of theoreticians such as Kekulé. William Henry Perkin (1838–1907), working at the age of 18 in the laboratory of August Wilhelm von Hofmann at the Royal College of Science in London, had been put to work on the synthesis of the drug quinine from aniline, the latter a coal-tar product that had been isolated by Hofmann. Perkin failed to synthesize quinine, but as a result of his efforts his flasks and just about everything else in his laboratory consistently ended up stained with a purplish substance. Perkin called this substance 'aniline purple', and when the French found it an excellent material for dyeing fabrics they named it mauve. The color became an immensely successful commercial product, and its widespread use created the 'Mauve Decade' of the 19th century. At the time

---

[3] Kekulé's other major contribution was his hypothesis, later shown to be correct in essentials though not in details, regarding the structure of the petroleum hydrocarbon known as benzene. This important chemical has the molecular formula C_6H_6, and it was Kekulé who first recognized, in 1865, that the six carbon atoms link to each other to form a ring:

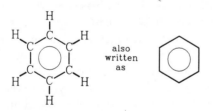

Kekulé reported his discovery as follows:

I was sitting, writing at my text-book; but the work did not progress, my thoughts were elsewhere. I turned my chair to the fire and dozed. Again the atoms were gambolling before my eyes. This time the smaller groups kept modestly in the background. My mental eye, rendered more acute by repeated visions of the kind, could now distinguish larger structures, of manifold conformation: long rows, sometimes more closely fitted together; all twining and twisting in snake-like motion. But look! What was that? One of the snakes had seized hold of its own tail, and the form whirled mockingly before my eyes. As if by a flash of lightning I awoke; and this time also I spent the rest of the night in working out the consequences of the hypothesis. Later he adds 'Let us learn to dream, gentlemen, then perhaps we shall find the truth . . . but let us beware of publishing our dreams before they have been put to the proof by the waking understanding.'

A snake biting its own tail seems a far remove from structural theory, but how many scientific advances arise in just such ways?

of Perkin's discovery (which was by no means the result of a planned, systematic synthesis based on an understanding of structural theory) dyes were derived from natural sources. It wasn't long after the young scientist's discovery that the synthetic dye industry was born; it was the first major industry based upon the science of organic synthesis, and it flourished, especially in Germany.

These two events, the solidification of structural theory and the discovery of a useful and commercially viable synthetic chemical, gave birth to the modern chemical world – the world that is much of the subject of this book. Industrial applications have led to the introduction into commerce of tens of thousands of organic chemicals, unknown to the world prior to the 1870s, and the science of toxicology grew in response to the need to understand how these new substances might affect the health of workers involved in their production and use, and of the rest of the population that might be exposed to them. It is of interest then to sketch a portrait of the organic chemical industry.

Industrial organic chemistry

Laboratory syntheses of organic chemicals proceeded at an astounding pace following the introduction and success of structural theory, and further impetus was provided by the economic success of the synthetic dye industry. Chemists have learned thousands of different ways to manipulate groups of atoms in organic compounds to create new molecular arrangements, and have also found how to develop sequences of individual chemical transformations that could lead to molecules of desired structural arrangement. They have learned how to lay out on paper a 'blueprint' for creating a molecule of specific structure and how to achieve that plan in the laboratory (although, of course, many plans fail to be achieved). If the synthesis is successful and the product useful, chemical engineers are called in to move the laboratory synthesis to the industrial production stage. Tens of thousands of synthetic organic chemicals, and similarly large numbers of naturally-occurring chemicals, are in commercial production.

In the late 19th century and up to World War II coal was the major 'starting material' for the organic chemical industry. When coal is heated in the absence of oxygen, coke and volatile by-products called coal tars are created. All sorts of organic chemicals can be isolated from coal tar – benzene, toluene, xylenes, ethylbenzene, naphthalene, creosotes, and many others (including Hofmann and Perkin's aniline). The organic chemical industry also draws upon other natural products, such as animal fats and vegetable oils, and wood by-products.

The move to petroleum as a raw materials source for the organic chemical industry began to occur during the 1940s. Petrochemicals, as they are called, are now used to create thousands of useful industrial chemicals. The rate of commercial introduction of new chemicals shot up rapidly after World War II.

Some of these chemicals are used primarily as *solvents* for one purpose or another. Many of the hydrocarbons found in petroleum, such as gasoline and kerosene, and the individual chemicals that make up these two mixtures, are useful as non-aqueous solvents, capable of dissolving substances not readily soluble in water. Hydrocarbons are highly flammable, however, and chemists found that conversion of hydrocarbons to chlorine-containing substances, using reactions in which hydrogen atoms were replaced by chlorine (below), create solvents still useful for dissolving substances not soluble in water, but having considerably reduced flammability. In the 1940s and 1950s, solvents such as these came into very wide use in the organic chemical industry and in many other industrial settings where they were needed (for degreasing of oily machinery parts, for example, or for 'dry' cleaning of clothing). As is now commonly acknowledged, we get no 'free lunch'. The so-called 'chlorinated hydrocarbon' solvents tend to be more toxic and more persistent in the environment than are the hydrocarbons. Reduced risks of fires and explosions bring increased risks of environmental harm. At the time industrial decisions were taken to move to the less flammable solvents, these kinds of trade-offs were little discussed; corporations are still learning how to balance risks and benefits, but now are at least aware that decisions of these types should not proceed without appreciable understanding of their environmental consequences.

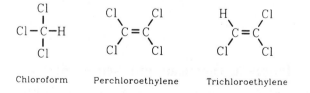

Chloroform   Perchloroethylene   Trichloroethylene

Among the thousands of other products produced by the organic chemical industry and related industries are included, in no particular order, medicines (most of which are organic chemicals of considerable complexity), dyes, agricultural chemicals including substances used to eliminate pests (insecticides, fungicides, herbicides, rodenticides and other 'cides'), soaps and detergents, synthetic fibers and rubbers, paper chemicals, plastics and resins of great variety, adhesives, substances used in the processing, preservation, and treatment of foods (food additives), additives for drinking water, refrigerants, explosives, cleaning and polishing materials, cosmetics, textile chemicals.

People can be exposed to greater or lesser degrees to most of these chemicals, indeed human exposures for many, such as medicines, are intended; and some people are exposed to other chemicals used in or resulting from production – the starting materials, the so-called 'intermediates' that arise during synthesis but which are not in the final products, and by-products of their production, including solvents and contaminants. The rate of development of new chemicals accelerated greatly after World War II, and this trend continues. Cesare Maltoni, a prominent figure in research on chemical carcinogens who directs The Institute of Oncology in Bologna, and Irving J. Selikoff of Mount Sinai School of Medicine in New York, whose work uncovering the link between asbestos exposure and human cancers is widely known, added this interesting footnote to their prefatory remarks at a major 1985 conference on chemical carcinogens held by the Collegium Ramazzini, in Bologna:

It is sobering to remember that the first man-made industrial chemical was created only in 1856, by Perkin in London. Factories to produce this and other aniline dyes were established in the next decades. The first bladder cancers associated with such dyes were reported by Rehn in 1895. At present the American Chemical Society is registering new chemicals at the rate of 70 *per hour*; some 500 enter commerce each year.

The organic chemical industry is vast and continues to grow. Human exposures to its products and by-products, whether intended or unintended, is one source of the chemical exposures we cover in this book.

Inorganic chemicals and their production

The history of man's efforts to tap the inorganic earth for useful materials is complex and involves a blend of chemical, mining, and materials technologies. Here we include everything from the various silicaceous materials derived from stone – glasses, ceramics, clays, asbestos – to the vast number of metals derived from ores that have been mined and processed – iron, copper, nickel, cadmium, molybdenum, mercury, lead, silver, gold, platinum, tin, aluminum, uranium, cobalt, chromium, germanium, iridium, cerium, palladium, manganese, zinc, etc., etc., etc. Other, non-metallic materials such as chlorine and bromine, salt (sodium chloride), limestone (calcium carbonate), sulfuric acid, and phosphates, and various compounds of the metals, have hundreds of different uses, as strictly industrial chemicals and as products consumers use directly. These inorganic substances reach, enter, and move about our environment, and we come into contact with them, sometimes intentionally, sometimes inadvertently. As with the organic chemicals, the subjects of this book are the potential health consequences of these contacts and exposures. Let us now see how these exposures are created.

2

Exposures

Our survival and that of all plants and animals on our planet depends upon chemicals: water, the nutrients in our foods, oxygen in our air, and if you are a plant, carbon dioxide as well. Oxygen, carbon dioxide, and water are simple inorganic chemicals, and the major nutrients – proteins, carbohydrates, fats, and vitamins – are organic compounds of considerable complexity. Certain inorganic minerals – calcium, zinc, iron, copper and a few others – are also essential to life. Living organisms can be envisioned as complex, highly organized collections of chemicals that absorb other chemicals from the environment and process them in ways that generate and store the energy necessary for survival of the individual and the species, primarily by the making and breaking of chemical bonds. It is a process of staggering complexity and beauty, and reveals in intricate detail the extraordinary inter-dependence of living organisms and their environments that has been created by the processes of evolution.

To maintain health, human beings and all other life forms must ingest, inhale, and otherwise absorb the essential environmental chemicals only within certain limits. If the amounts we take into ourselves fall below a certain level, we may suffer malnutrition or dehydration, or suffer the effects of oxygen deprivation. If we take in too much, we become obese, develop certain forms of cancer, heart disease, and many other diseases as well. We can even poison ourselves by consuming large amounts of essential vitamins and minerals. Living organisms can certainly tolerate a fairly wide range of intake of these essential chemicals, but there are limits for all of them beyond which we get into trouble. Scientists are still learning a great deal about what those limits are, and as more is learned we shall no doubt be able to put this knowledge to use to reduce the risks of a number of important diseases and to enhance the health and well-being of all life.

When we take a drink of water we expose ourselves not only to molecules of H_2O, but other chemicals as well. Depending upon the source of water, we are typically ingesting a variety of minerals, some of which are essential to health, but many of which have no known role in preserving our health. We are also consuming some organic chemicals that have migrated from plants or soil microorganisms. These minerals and organic chemicals are naturally-occurring, but in many areas, drinking water will also contain certain minerals at concentrations in excess of natural levels because of some type of human activity – mining, manufacturing, agriculture – and synthetic organic chemicals that have somehow escaped into the environment. We also intentionally add some chemicals to water to achieve certain technical effects, including the very important effect of disinfection.

The situation is the same with the air we breathe. We need oxygen, and so cannot avoid nitrogen, carbon dioxide and several other naturally-occurring gases. We are also inhaling with every breath a variety of naturally-occurring and industrial chemicals that are either gases or that are liquids that are volatile enough to enter the gaseous state. The specific chemicals and their amounts depend on where we are when we are breathing.

The plants and animals we have chosen to use as foods naturally contain thousands of chemicals that have no nutritional role, and when we eat to acquire the nutritionally essential chemicals we are automatically exposed to this huge, mostly organic, chemical reservoir. Of course, human beings have always manipulated foods to preserve them or to make them more palatable. Processes of food preservation, such as smoking, the numerous ways we have to cook and otherwise prepare food for consumption, and the age-old methods of fermentation used to make bread, alcoholic beverages, cheeses and other foods, cause many complex chemical changes to take place, and so result in the introduction of uncounted numbers of compounds that are not present in raw agricultural products.

Human beings have for many centuries been quite good at manipulating the genetic material of food plants and animals to produce varieties and hybrids with improved characteristics of one type or another. For most of the time breeders had no idea what they were doing, because little was known about genetics and the consequences of manipulating genetic material. Now we know that when we create a new tomato hybrid we are producing changes in the chemical composition of the fruit; from the chemist's point of view, a tomato is not a tomato is not a tomato. People eating the varieties of tomatoes available 50 years ago did not ingest exactly the same collection of compounds found in varieties available nowadays.

Spices and herbs contain no nutritionally essential chemicals of consequence, but they do contain hundreds of organic compounds that impart flavors, aromas, and colors. Here, people deliberately expose themselves to a plethora of chemicals largely for aesthetic reasons.

Like air and water, foods may also become contaminated with man-made chemicals, and certain unwanted, naturally-occurring substances such as aflatoxins. For centuries, certain chemicals have been added to food to achieve a variety of technical effects.

We also paint our bodies with all sorts of colors and splash perfumes and other cosmetics on our skin. We wash ourselves and our environments with chemicals. We take medicines, many of which are exceedingly complex organic molecules. We use chemicals to rid ourselves and our foods of pests. Our bodies contact materials – chemicals – we use for clothing and to color clothing. Most of us are exposed to chemicals on the job and through the hundreds of products we use in the house and for recreation. Fuels and products resulting from their combustion and materials used for buildings add to the burden. In the United States 30 million people can't seem to avoid the several thousand chemicals they inhale after every puff of a cigarette.

Food, air, water, consumer products, cosmetics, pesticides, medicines, building materials, clothing, fuels, tobacco products, materials encountered on the job, and unwanted contaminants of all of these. These are the principal sources of the thousands, perhaps hundreds of thousands of known and unknown chemicals, natural and man-made, that people ingest, inhale, absorb through their skin, and (for medicines) take into their bodies in other ways. We should also mention unusual, but not uncommon, forms of exposure: venoms and other substances from animals that may bite or sting us and plants with which we may come into contact, and soils and dusts we inadvertently or intentionally ingest or inhale.

As with the very small fraction of these chemicals that are essential to our health, it appears there are ranges of exposures to all these chemicals that, while probably not beneficial to health, are probably without detrimental effect. And, for all of them, both the man-made ones and those of strictly natural origin, there are ranges of exposure that can put our health into jeopardy. These facts bring out one of the most important concepts in toxicology: all chemicals are *toxic* under some conditions of exposure. What the toxicologist would like to know are those conditions. Once they are known, measures can be taken to limit human exposures so that toxicity can be avoided.

In the next chapter, and for a large part of the book, toxicity will be the main topic. For now we need only note that by *toxicity* is meant the production of any type of damage, permanent or impermanent, to the structure or functioning of any part of the body.

A more complete elaboration of the technical issues and terms related to chemical exposures is desirable, and is the subject of the remaining sections of this chapter.

Environmental media

Chemicals reach us through various *media*. By media is meant the vehicles that carry the chemical and that get it into contact with the body. Thus, diet, air, water, and soils and dusts are the principal environmental media through which chemical exposures take place. Direct contact with the chemical, as with cosmetics applied to the skin or household products accidentally splashed into the eye, may also occur, in which case the cosmetic or household product may be said to be the medium through which exposure occurs.

Exposure pathway

It is often of interest to understand the *pathway* a chemical traveled to reach the medium that ultimately creates the exposure. This typically takes the form of a description of the movement of a chemical through various environmental media. Some pathways are short and simple: aflatoxin, for example, contaminates a moldy food and we consume that food; there is only one medium (the food) through which the aflatoxin moved to reach us. Most pathways are somewhat more complex.

Lead added to gasoline (medium 1) is emitted to the air (medium 2) when gasoline is burned. Some of the airborne lead deposits in soil (medium 3) which is used for growing corn. Some of the lead in soil dissolves in water (medium 4) and moves through the roots of the corn plant, accumulating in the kernels of corn (medium 5). The corn is fed to dairy cattle and some of the lead is excreted in milk (medium 6). Milk is the medium that creates human exposure to the lead. The lead has passed through six media to reach a human being. To make matters more complex, note that people may be exposed to lead at several other points along the pathway, for example by breathing the air (medium 2) or coming into contact with the soil (medium 3).

Knowledge of exposure pathways is a critical part of the analysis needed to piece together the human exposure pattern.

Exposure routes

The *route* of exposure refers to the way the chemical moves from the exposure medium into the body. For chemicals in the environment the three major routes are *ingestion* (the oral route), inhalation, and *skin contact* (or *dermal* contact). Medicines get into the body in these three ways, but in several other ways as well, for example by injection under the skin or directly into the bloodstream, or by application to the eye.

In many cases, a given medium results in only one route of exposure. If diet is the medium, then ingestion will be the exposure route. If air is the medium, the chemical enters the body by inhalation. Immediately, however, we can think of ways this simple rule does not hold. Suppose, for example, that a chemical is contained in very small particles of dust that are suspended in air. The air is inhaled and the dust particles containing the chemical enter the airways and the lungs. But some of these dust particles are trapped before they get to the lungs and others are raised from the lungs by a physiological process to be discussed later. These particles can be collected in the mouth and swallowed. So here is an example of a single medium (air) giving rise to two exposure routes (inhalation and ingestion). These types of possibilities need to be considered when exposures are being evaluated.

The dose

To determine whether and to what extent humans may be harmed (suffer toxicity) from a chemical exposure, it is necessary to know the *dose* created by the exposure. The concept of dose is so important that it needs to be treated in detail.

Everyone is generally familiar with the term dose, or dosage, as it is used to describe the use of medicines. A single tablet of regular strength aspirin typically contains 325 milligrams (mg) of the drug. An adult takes four tablets in one day, by mouth. The total *weight* of aspirin ingested on that day is 1300 mg, or 1.3 grams (1 mg = 0.001 g). But weight is not dose. For reasons relating to how aspirin affects the body biologically, i.e., how it relieves pain, the critical measure is the amount taken into the body divided by the weight of the person, expressed in kilograms (1 kg = 1000 g). The aspirin *dose* for a 65 kg adult (about 145 lbs) is thus 1300 mg/65 kg = 20 mg/kg body weight (b.w.). The time over which the drug was taken is also important in judging its effectiveness. Our adult took four tablets in one day, and the day is the usual time unit of interest. So a more complete description of the aspirin dose in this case is given by 20 mg/kg b.w./day. The typical *dose units* are thus milligram of chemical per kilogram of body weight per day (mg/kg b.w./ day).

Note – and this is quite important – that if a 20 kg child (about 44 lbs) were to take the same four tablets on one day, his dose would be more than three times that of the adult, as follows: 1300 mg aspirin/20 kg b.w. = 65 mg/kg b.w./day. For the same intake of aspirin, the lighter person receives the greater dose.

Calculating doses for environmental chemicals is not much more complex. Suppose a local ground water supply has become contaminated with the widely used degreasing solvent, trichloroethylene (TCE). EPA scientists have measured the extent of TCE contamination and found that the water contains 2 micrograms of TCE in each liter of water (how they know this will be discussed a little later). People living above the water have sunk wells, and adults are drinking two liters of the water each day and their children are consuming one liter.

First we calculate the *weight* of TCE (in mg) getting into their bodies.

Adults
Consuming 2 liters per day of water containing 2 micrograms of TCE in each liter leads to ingestion of 4 micrograms of TCE each day.
A microgram is 0.001 mg.
Therefore, weight of TCE taken into the body is 0.004 mg/day.

Children
Consuming 1 liter per day of the same water leads to an intake of 0.002 mg/ day.

If the adults weigh 80 kg and the children weigh 10 kg, then their respective daily *doses* are:

Adults
0.004 mg per day/80 kg b.w. = 0.00005 mg/kg b.w./day

Children
0.0020 mg per day/10 kg b.w. = 0.0002 mg/kg b.w./day.

The child again receives a higher dose (4 times) than the adult consuming the same water, even though the child consumes less water.

Ultimately it must be asked whether a health risk exists at these doses, and a second important factor in making this determination is the number of days over which the dose continues. *Duration* of exposure as well as the dose received from it thus needs to be included in the equation.

To be sure of our terms: dose and its duration are the critical determinants of the potential for toxicity. Exposure creates the dose. In our example, people are *exposed* to TCE through the *medium* of their drinking water, and receive a *dose* of TCE by the oral *route*.

Calculating (or, more correctly, estimating) dose requires knowledge of the weight of the chemical getting into the body by each route. As in the account just given of exposure to a chemical in drinking water, the amount of chemical in a specified volume of water (mg/l) times the amount of that water consumed by a person each day (l/day) gave the weight of chemical taken into the body each day. If, instead of drinking water, the chemical is in air, then the required data are weight of chemical for a given volume of air (usually mg per cubic meter, m^3) and the volume of air a person breathes each exposure period (m^3 per day). Suppose the air in a gasoline station contains 2 mg carbon monoxide per cubic meter and a worker breathes that air for an 8 hour work day. Typically, an adult engaged in moderate activity will breathe in 10 m^3 of air in 8 hours. Thus 2 mg/m^3 × 10 m^3/8 hrs = 20 mg/8 hrs. If the worker does not inhale carbon monoxide for the rest of the day away from the station, then the daily intake is 20 mg. In fact, he will probably also inhale smaller amounts for the rest of the day from other sources and thus receive a somewhat higher total daily dose.

Food intakes are estimated in the same way. The amount of chemical per unit weight of food is multiplied by the daily intake of that food item to obtain the weight of chemical ingested each day. Dose is then calculated by dividing the weight of the chemical by the body weight of the exposed person. Estimating food intake is not quite so straightforward as estimating air volumes inhaled or water volumes consumed. Peoples' intakes of different foods vary greatly. If we need to estimate the amount of the pesticide Alar someone might have received (prior to 1989, when its use ended) from residues on apples, it is critical to understand whether we are interested in the average eater of apples (40 g per day, or about 1.3 ounces) or above average eaters. Ten per cent of the U.S. population consumes, for example, more than 140 g of apples per day (about one-third of a pound), and these people would obviously have been exposed to a greater dose of the pesticide than the average eater, assuming all apples had the same amount of Alar (which they did not). Toxicologists are generally concerned to ensure that risks to individuals exposed to higher than average amounts of chemicals are avoided, although highly eccentric eating patterns are very difficult to take into account.[4]

Information on food consumption rates and their distribution in the population is available, but it is far from perfect. Because peoples' food consumption habits change over time, and because so much eating now takes place in restaurants and from foods prepared outside the home, it is becoming increasingly difficult to acquire reliable data on this subject. As with so many other aspects of the science of toxicology and risk assessment, the needed data are either lacking or highly uncertain. Scientists in this field are under increasing pressure to supply firm answers without the benefit of firm data.

Concentrations

Several references have been made to the amount of chemical in a given amount of medium. The technical term that describes this relationship is *concentration*, sometimes referred to as *level*. The concentrations of chemicals are typically expressed in different, though similar, units, according to the medium:

Medium	Concentration units
water	milligram chemical/liter water (mg/l)
food	milligram chemical/kilogram food (mg/kg)
air	milligram chemical/cubic meter air (mg/m^3)

A less precise but widely used unit is that of 'parts per'. Because such units are often used by TV and the press, and in other public presentations of information on chemicals in the environment, they shall be described here.

---

[4] In a survey conducted at the time of the saccharin scare in 1976, a tiny portion of the U.S. population was found to consume more than 36 cans of diet soda each day!

For water, 'parts per' refers to parts of chemical per so many parts of water. If there are 10 milligrams of TCE in one liter of water, then the concentration is said to be 10 parts-per-million (ppm). How does this come about? First, it must be recognized that one liter of water weighs exactly one kilogram. One milligram is one one-thousandth of one gram. One kilogram is 1000 grams. Thus, a milligram is *one one-millionth* of one kilogram. So, one mg TCE per kg water is also one part per million. We then see that 10 mg/l is equivalent to 10 ppm.

If instead of mg of the chemical, it is found to be present at a concentration of micrograms/liter, then the equivalent units are parts-per-billion (ppb), because a microgram is one-billionth of a kilogram. Because these units will come up throughout the book, a reference table on them will come in handy.

Some measures of weight and volume: metric system

Weight

kilogram (kg) (1000g)	(approx 2.2 lbs)
gram (g)	(approx 1/30 oz)
milligram (mg) (0.001g)	
microgram (μg) (0.000 001g)	
nanogram (ng) (0.000 000 001g)	
picogram (pg) (0.000 000 000 001g)	

Volume

liter (l)	(approx 0.9 U.S. quart)
milliliter (ml)	(0.001 liter)
cubic meter (m³)	(1000 liters)

The concentrations of chemicals in environmental media can be known in two major ways. They can be known either because the chemical has been added to known amounts of a medium in known quantities, or because the concentrations have been measured using any of several technologies developed and applied by analytical chemists.

Certain substances are deliberately added to food to achieve desired technical effects: to preserve, to color, to stabilize, to emulsify, to sweeten, and so on. Such substances are added under carefully controlled conditions so that the concentrations of additives in food are generally known with a high degree of accuracy. Pharmaceuticals are similarly packaged and so concentrations of these substances are also known with high accuracy.

The situation is not so simple with most agents in the environment, and concentrations have to be measured. All measurements have limitations that have to be recognized.

Suppose a pesticide is applied to tomatoes to control fungal invasion. To gain approval for such a pesticide use in the United States, the manufacturer would have had to conduct and submit studies to the EPA regarding the amount of pesticide (in this case a fungicide) to be applied to insure effective fungal control, and the amount of the fungicide that would remain on the tomatoes at harvest time. If EPA judged such pesticide residue concentrations to be safe (actually, judgment is made on the dose a person would receive from consuming the tomatoes with the specified residue concentration), the manufacturer could receive approval to market the fungicide. This approval is called a pesticide registration. The marketed product would be required to carry a label specifying the approved rate of application, so that the use of the fungicide would result in residue concentrations no greater than those considered by EPA to be safe.

Knowledge of the concentrations of the fungicide in tomatoes and tomato products depends in this case on measurement. In theory, one might estimate the amount of fungicide on a tomato by calculation, based on the rate of application to the crop. But this is for several reasons an extremely uncertain calculation. To calculate accurately it is necessary to know how much of the applied chemical is on the actual tomatoes as against the remaining parts of the plant. It is also necessary to understand how much might be lost by certain physical processes – rain, wind – as the tomato grows. The rate of chemical degradation of the fungicide has to be understood. The extent to which the fungicide becomes concentrated or diluted when tomatoes are processed to juice, paste, ketchup, and other products needs to be known. It is clearly much easier and more reliable to have the analytical chemist measure the concentrations of residual fungicide in the tomatoes and tomato products than to estimate residue levels by calculation.

Such analytical measurements are necessary to establish concentrations for most agents in the environment. How much benzene is present in the air at gasoline stations as a result of its evaporation from gasoline? What is the concentration of arsenic in water running off the surface of a hazardous waste site where unknown amounts of arsenic were buried over many years? What is the polychlorinated biphenyl (PCB) concentration in fish swimming in waters next to a hazardous waste site known to contain this substance? How much aflatoxin is in a lot of peanut butter? The most reliable answers to these questions are those resulting from chemical analysis.

Analytical chemistry has undergone extraordinary advances over the past two to three decades. Chemists are able to measure many chemicals at the part-per-billion level which in the 1960s could be measured only at the part-per-million level (that's 1000 times more concentrated), or even the part-per-thousand level. For a few chemicals measurement science reaches to the part-per-trillion level. (Even more dilute concentrations can be measured for some substances, but not for those in the environment.) These advances in detection capabilities have revealed that industrial chemicals are more widespread in the environment than might have been guessed 10 or 20 years ago, simply because chemists are now capable of measuring concentrations that could not be detected with analytical technology available in the 1960s. This trend will no doubt continue, because analytical chemists do not like to be told to stop developing the technology needed to search for smaller and smaller concentrations of chemicals: 'chasing zero' is a challenge chemists are trained to pursue, even though they all know 'zero' can never be reached.

While analytical science is in many ways quite miraculous, it is by no means without problems. Errors are easily made. Analyses are not always readily reproducible in different laboratories. Some technologies are exceedingly expensive. And while analytical methods are well worked out for many chemicals, they are not available at all for many more. (Indeed, if we are interested in the naturally occurring chemicals that human beings are exposed to, we will find that only a tiny fraction of these can now be analyzed for with anything except fairly sophisticated research tools; most such chemicals are still unknown, and it is virtually impossible to develop routine and reliable analytical methods for chemicals that have not even been characterized.)

A third way to gain some knowledge about the concentrations of chemicals in the environment involves some type of modeling. Scientists have had, for example, fair success in estimating the concentrations of chemicals in the air in the vicinity of facilities that emit those chemicals. Information on the amount of chemical emitted per unit time can be inserted into various mathematical models that have been designed to represent the physical phenomena governing dispersion of the chemical from its source. Certain properties of the chemical and of the atmosphere it enters, together with data on local weather conditions, are combined in these models to yield desired estimates of chemical concentrations at various distances from the source. These models can be 'calibrated' with actual measurement data for a few chemicals, and then used for others when measurement data are not available.

Sampling the environment

Another uncertainty in understanding environmental concentrations arises because of the problem referred to as *sampling*. Suppose inspectors from FDA want to know whether a shipment of thousands of heads of lettuce from Mexico contains illegal concentrations of a particular pesticide. Obviously, the entire shipment cannot be sampled, because analysis destroys the lettuce. So the inspector takes a few heads from different areas of the shipment and these are either combined in the laboratory (a 'composite' analysis) or analyzed individually. In either case a certain concentration (for the composite) or a range of concentrations (for the individual heads) is reported from the laboratory. How can the inspector be sure these results fairly represent the entire lot?

Suppose EPA scientists are investigating lead contamination of soil near an abandoned mining and smelting operation. How should the soil be sampled so that the analytical results obtained on the individual samples provide a reliable indication of the range and distribution of contamination of the entire site?

In neither the case of the pesticide residues in lettuce nor the lead concentrations in soil can it be certain that the whole is accurately represented by the part taken for analysis. Scientists can only have various degrees of confidence that the samples taken are representative. Statisticians can devise sampling plans that, if followed, allow scientists to know the degree of confidence, but that's the best that can be done. In practice much environmental sampling is done without a well-thought-out plan, in which case no statement can be made about the degree to which the whole is represented by the part. This unfortunate and unscientific practice complicates the lives of decision-makers, who many times have no recourse but to ignore the problem. (There are occasions when 'statistical representativeness' is not very important. If we suspect the presence of some serious and potentially life threatening contamination we want to learn about it as quickly as possible; here we sample not to understand everything, but to learn whether some emergency action is needed.)

The scientific issues associated with understanding and estimating human exposures to chemicals in the environment are vastly more complicated than has been suggested in this chapter, but for present purposes they have been covered sufficiently. The primary purpose of the discussion is to introduce some terms that will come up frequently in later chapters, and to provide some insight regarding how scientists come to understand how much of which chemicals are present in environmental media, and some of the ways they can come to enter the human body. At this point we are coming close to the central topic of toxicity, which obviously cannot occur until chemicals actually contact various parts of the body. But there is one more step that needs to be examined before toxic effects are considered. How do chemicals enter, move around within, and exit the body?

3

Into the Body

Chemicals in the environment may enter the mouth and be swallowed into the gastrointestinal tract. If they are in the vapor state or are attached to very fine dusts in the air, they may be inhaled through the nose and mouth and thereby enter the airways leading to the lungs. Some chemicals reach the skin, sometimes dissolved in some medium, sometimes not. What happens following the contact of environmental chemicals with these three routes of entry to the body?

First, chemicals come into intimate contact with the fluids, tissues, and cells that make up these three passages into the body. This contact may or may not result in some type of injury to tissues and cells; if some adverse response occurs in the tissues comprising these entryways, it is referred to as *local* toxicity.

In most cases, however, chemicals enter the bloodstream after they are *absorbed* through the walls of the gastrointestinal tract or the lungs, or through the layers that make up the skin. Once in the bloodstream they can be *distributed* through the body and reach the tissues and cells that make up the many organs and systems of the body. In most cases chemicals undergo molecular changes – chemical reactions in the cells of the body's organs, particularly the liver but in others as well: they are *metabolized*. Metabolism is brought about by enzymes – large protein molecules – that are present in cells. Chemicals and their metabolites (the products of metabolism) are then *excreted* from the body, typically in urine, often in feces and in exhaled air, and sometimes through the sweat and saliva.

The nature of toxic damage produced by a chemical, the part of the body where that damage occurs, the severity of the damage, and the likelihood that the damage can be reversed, all depend upon the processes of absorption, distribution, metabolism and excretion, ADME for short. The combined effects of these processes determine the concentration a particular chemical (the chemical entering the body or one or more of its chemical reaction products, or metabolites) will achieve in various tissues and cells of the body and the duration of time it spends there. Chemical form, concentration, and duration in turn determine the nature and extent of injury produced. Injury produced after absorption is referred to as *systemic toxicity,* to contrast it with local toxicity. It is worthwhile to spend some time to understand ADME.

First, a word about how ADME is typically studied. A highly interesting technique is used to follow a chemical through the body. Chemists have learned to increase the natural level of certain radioactive forms, or isotopes, of carbon, hydrogen, and some other atoms in organic compounds. They can, for example, create molecules enriched in carbon-14, a radioactive isotope having two more neutrons in the atom's nucleus than does the most abundant, non-radioactive isotope of carbon. The ADME pattern of the radioactively-labelled chemical can be easily traced – indeed, these are called radioactive tracer studies – because it is relatively simple to locate in the body the extra radioactivity associated with the added carbon-14. The presence of extra amounts of the radioactive isotope of carbon alters in no significant way the chemical behavior of the molecule carrying it. Radiotracer studies have not only made possible modern toxicology, but almost all of modern biochemistry and pharmacology depend upon the use of this technique. Its use is limited to experimental systems because, except for medical treatment, extra radioactivity cannot be administered to people; but this is nevertheless immensely beneficial.

Those substances in the environment that are essential to the health of the organism – the nutrients, oxygen, water – all undergo ADME. Everyone is familiar with the fact that carbohydrates, proteins and fats break down in the body – the initial stages occurring in the gastrointestinal tract, and the rest in the liver and other organs after absorption – and that the chemical changes these essential chemicals undergo (their metabolism) are linked to the production and storage of energy and the synthesis of the molecules that are essential to life. Everyone is also familiar with the fact that there are variations among people in the ease with which they can absorb certain nutrients – iron and calcium, for example – from the gastrointestinal tract into the blood. Although the chemicals we shall be discussing are not essential to life, and can be quite harmful under some conditions, our bodies have mechanisms for absorbing, distributing, metabolizing, and excreting them not greatly different from the ways our bodies handle essential nutrients.

Absorption

Gastrointestinal tract

Figure 1 depicts in schematic form the relationships among certain organs and systems of the body essential for an understanding of ADME. The arrows in the Figure depict the various paths chemicals follow when they enter the body, move around within it, and are finally excreted from it. It will be helpful to refer to the Figure throughout the course of this chapter.

It is not an oversimplification to describe the gastrointestinal (GI) tract as one very long tube open at both ends, the mouth and the anus. The major pieces of the tube are the mouth, throat, esophagus, stomach, small intestine and large intestine, or colon, and the rectum and anus. Most of the length is due to the intestines, which are actually highly coiled.

Chemicals in food and water, some medicines, and even some present in soils or dusts that are incidentally ingested are absorbed along the entire GI tract. By absorption is meant the movement of the chemical through the membranes of the different types of cells comprising the wall of the GI tract so that it ends up in the bloodstream.

There are several different biological mechanisms at work to affect absorption, but their nature need not concern us here. Suffice it to say that a host of factors affects the site along the GI tract where a chemical is absorbed and the rate and extent of its absorption. Among these are the particular chemical and physical properties of the chemical itself, the characteristics of the medium – food (even the type of food) or water – in which it enters the GI tract, and several factors related to the physiological characteristics of the exposed individual. Toxicologists refer to the latter as *host* factors, because they belong to the individual that is playing 'host' to the entering chemical.

Certain drugs are administered as sublingual tablets (they are placed under the tongue) and as rectal suppositories; these substances can be

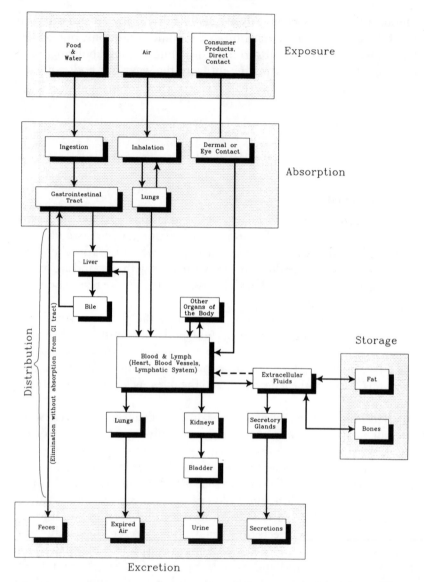

Figure 1. A schematic showing how chemicals may enter, be absorbed into, distributed within, and excreted from the body. Adapted in part from C.D. Klaassen, Distribution, Excretion, and Absorption of Toxicants, Chapter 3 of Casarett and Doull's Toxicology *(see Recommended Readings).*

absorbed in the mouth and rectum, respectively. These two GI tract sites appear to be minor routes of entry for most environmental chemicals; the latter are more typically absorbed through the walls of the stomach and intestines.

Chemicals vary greatly in the extent to which they are absorbed through the walls of the GI tract. At one extreme are some very inert and highly insoluble substances – sand (silicon dioxide) and certain insoluble minerals such as several of the silicates added to foods to keep them dry – that are almost entirely unabsorbed. Such substances simply wind their way down the entire length of the GI tract and end up excreted in feces. This pathway is shown in Figure 1 as the long arrow extending from the GI tract directly to feces.

Most substances are absorbed to a degree, but few are entirely absorbed. Lead absorption from food, for example, may be in the range of 50%, but is less when this heavy metal is in certain highly insoluble chemical forms or when it is associated with certain media such as dried paint or soils.

Much else is known about GI absorption. Individuals vary in the extent to which they can absorb the same chemical, and absorption can be influenced by individual factors such as age, sex, health status, and even dietary habits. People who consume large amounts of fiber may absorb less calcium and iron than those who eat less. The GI tract is not fully developed until about 24 months after birth, and infants absorb metals such as lead and certain organic chemicals more readily than do adults.

Different animal species exhibit differences in GI absorption rates. The extent of GI absorption of lead in rats, for example, can be studied by feeding the animals known amounts of the metal and analyzing the unabsorbed amount that comes through in feces; the difference is the amount absorbed. But because of possible species' differences it is not possible to conclude that humans will absorb the same amount of lead as the rat. These types of differences complicate evaluation of toxic potential. At the same time, they help to explain why different species of animals respond differently to the same dose of a chemical.

Respiratory tract

The respiratory tract includes the air passages through the nose and mouth that connect to the tubing (bronchi) that lead to the lungs. Gases such as oxygen and carbon dioxide, and the environmentally important pollutants carbon monoxide, nitrogen oxides, sulfur dioxide and ozone, can readily travel the length of the respiratory tract and enter the lungs. So can vapors of volatile liquids such as gasoline and certain solvents. Sometimes such chemicals cause local toxicity – everything from minor, reversible irritation of the airways to serious, irreversible injury such as lung cancer – but, as with the GI tract, some amounts of these agents pass through the lungs (in the so-called alveolar area) and into the blood. The rates and extent of lung absorption are influenced by a number of host factors, including physiological variabilities specific to the animal species and even to individuals within a species.

Dusts in the air can also enter the airways. Here the physical dimensions of the individual particles determine the degree to which they migrate down the respiratory tract and reach the lung. Generally, only very fine particles, those smaller than about one micron (one-millionth of a meter), enter the deep (alveolar) region of the lung. Larger particles either do not enter the respiratory tract or are trapped in the nose and excreted by blowing or sneezing. Some particles deposited in the upper regions of the respiratory tract may be carried to the pharynx and be coughed up or swallowed. Thus, inhaled chemicals or dusts can enter the body by the GI tract as well as the respiratory tract.

Dusts may carry chemicals into the lungs, where they can be absorbed by several mechanisms. But there are other physical materials – asbestos is the most well known – that, depending upon their physical dimensions, can also be inhaled and can move down the respiratory tract to the lung, where they can cause damage.

Skin

The skin acts as a barrier to the entry of chemicals, but some chemicals get through it. Dermal, or percutaneous absorption, as it is technically called, generally involves diffusion of a chemical through the so-called epidermis, which includes the outer layer of dead cells called the stratum corneum. This is a tough barrier for chemicals to get through, and many don't make it. If they do, they also have to negotiate passage through the less protective second layer called the dermis; once by this they are into the blood.

The effectiveness of the stratum corneum in blocking the passage of chemicals varies from one part of the human body to another. It is particularly difficult for chemicals to cross the palms of the hands and soles of the feet, but they get by the scrotum fairly easily. Abdominal skin is of intermediate effectiveness in preventing absorption. Age and sex also influence rates of dermal adsorption, and some species of animals (e.g. rabbits) seem to have much more vulnerable skins than others; humans and other primates appear to be near the least vulnerable end of the scale.

Not surprisingly damage to the skin enhances absorption rates; if the less protective dermis is exposed because the stratum corneum has been scraped off, penetration can be substantial.

The physical properties of a chemical, which are in turn functions of its chemical structure, have a powerful influence on the likelihood it will get through the skin. Generally, chemicals must be capable of dissolving fairly readily in both water and fat-like materials. Substances that dissolve only in water and those that have little affinity for water but only for fatty materials, do not get far. Large molecules can not move as easily through the skin as can smaller ones. Substances that do not dissolve well in water or any other solvents just cannot penetrate in measurable amounts.

Distribution

Once absorption occurs and a chemical is in the blood it can move around the body with relative ease (see Figure 1), going just about everywhere blood goes. It generally will not be distributed equally to all organs and systems; and the pattern of distribution will vary greatly among chemicals, according to their particular structural characteristics and physical properties.

There are fortunately a few natural biological barriers that prevent or impede distribution of chemicals to certain organs. The most important of these are the *blood–brain* barrier and the *placental* barrier, the one retarding entry of chemicals to the brain, the other protecting the developing fetus. These barriers are not perfect, however, and certain chemicals can migrate through them. Most chemical forms of the metal mercury, for example, can not readily pass the blood–brain barrier, and they exert their primary toxic effects on the kidney, not the brain. But there is a certain chemical form of mercury, called methyl mercury, that can break through the barrier, mostly because it can dissolve in fatty materials while the other forms of mercury can not, and this form can cause damage to the brain. Methyl mercury can also invade the placental barrier while other forms of mercury are largely locked out. Other biological barriers, none perfect, exist in the eye and testicles.

Certain chemicals can be *stored* in the body, as depicted in Figure 1. A major site of storage is bone, which can bind metals such as lead and strontium and non-metallic inorganic elements such as fluoride. While bound in this form the chemicals are relatively inert, but under certain conditions, they can be slowly released from storage and reenter the bloodstream where they are more available to cause biological effects.

Another tissue – fat – can store certain organic chemicals that are highly soluble in this medium. Certain pesticides such as DDT and industrial products such as polychlorinated biphenyls (PCBs) readily dissolve in body fat and can stay there for long periods of time. Most people have measurable amounts of these two once widely-used chemicals, and several more as well, in their fat stores.

Chemicals stored in fat and bone are at the same time also present in blood, usually at very much lower concentrations. A kind of equilibrium exists between the storage tissue and blood. If more of the chemical is absorbed into the body, its blood concentration will first rise, but then will fall as some of it enters the storage area; equilibrium conditions are eventually reestablished, with higher concentrations of the chemical in both the medium of storage and the blood. Likewise, removal of the source of exposure and loss of the chemical from blood through excretion (see below) will mobilize the chemical from its storage depot and send it into the blood. As long as no external sources of the chemical are available, it will continue to be lost from the blood and continue to migrate out of storage. Finally, it will all but disappear from the body, but this may take a very long time for some chemicals.

An interesting phenomenon has been observed in people who have lost weight. Removal of body fat decreases the amount available for storage of fat-soluble chemicals. Blood concentrations of DDT have been observed temporarily to increase following weight loss. In light of the discussion above, such an observation is not unexpected.

Metabolism

The cells of the body, particularly those of the liver, and with important contributions from those of the skin, lungs, intestines and kidneys, have the capacity to bring about chemical changes in a large number of the natural and synthetic chemicals that are not essential to life. As we have said, these chemical changes yield metabolites of the absorbed chemical, and the process whereby metabolites are produced is called metabolism. The latter term is also applied to the biochemical changes associated with nutrients and substances essential to life.

Often metabolites are more readily excreted from the body than the chemical that entered it, and chemical pathways leading to such metabolites are called detoxification pathways; the quicker the chemical is eliminated, the less chance it has to cause injury.

Metabolism is generally brought about, or catalyzed, by certain proteins called enzymes. Cells have many enzymes and most are involved in biochemical changes associated with the ordinary life processes of cells. But some act on non-essential chemicals and convert them to forms having reduced toxicity and enhanced capacity for excretion. Why we should have such enzymes is not clear, but some are probably the result of evolutionary adaptations that increased species survival in the face of environmental threats. These enzyme systems evolved over very long periods of time, primarily in response to millions of years of exposure of cells to naturally-occurring chemicals, and they were thus available for the exposures to synthetic chemicals that began only about 125 years ago. Most synthetic chemicals have the same groupings of atoms that are found in naturally-occurring molecules, although important differences certainly exist. In particular, the carbon–chlorine bond, common in some important industrial solvents and other chemicals, is relatively rare in nature. It is perhaps not surprising that so many chemicals that have prompted public health concerns contain carbon–chlorine bonds.

An example of beneficial metabolism is illustrated by the conversion of toluene, a volatile chemical present in petroleum products and readily absorbed through the lungs, to benzoic acid, as shown:

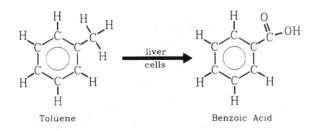

Toluene   Benzoic Acid

Toluene is chemically related to benzene (footnote, page 7), but a methyl (CH$_3$–) grouping has replaced one of the hydrogen atoms attached to the ring of six carbon atoms. The 'circle' drawn inside the six-carbon ring represents a set of six electrons that comprises the chemical bonds involved in the so-called aromatic ring. This collection of six carbon atoms in a ring with six electrons represented by the circle appears in many organic compounds, including a number appearing later in this book.

At high exposures toluene molecules can reach and impair the nervous system. But the liver has certain enzymes that can eliminate the three hydrogen atoms attached to the carbon atom, those of the methyl group, and introduce oxygen atoms in their places. The rest of the toluene molecule is unaffected. But this metabolic change is enough. Benzoic acid is much less toxic than toluene – indeed, it has very low toxicity – and it is much more readily excreted from the body. This metabolic pathway detoxifies toluene. Of course, if an individual inhales huge amounts, such that the amount of toluene absorbed and distributed exceeds the capacity of the liver to convert it to benzoic acid, then toxicity to the nervous system can be caused by the excess toluene. Though toluene is chemically related to benzene, the latter is a far more toxic chemical – it can cause certain blood disorders and forms of leukemia in humans – in part because it cannot so readily be metabolized to chemical forms having reduced toxicity.

Toxicity can occur because, unfortunately, some metabolites are, unlike benzoic acid, more toxic than the chemical that enters the body. Enzymes can cause certain changes in molecular arrangements that introduce groupings of atoms that can interact with components of cells in highly damaging ways. The industrial chemical bromobenzene can be converted in the liver to a metabolite called bromobenzene epoxide, as depicted in the diagram.

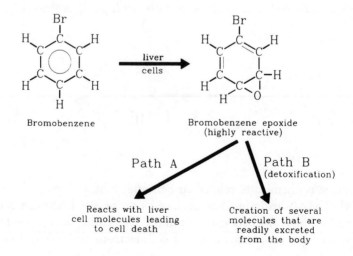

The epoxide molecule is very active and can chemically bind to certain liver cell molecules and cause damage and even death to the cell (Path A). But an alternate reaction path (B) can also operate. If the amount of bromobenzene that enters the cell is low enough, Path B (which actually creates several metabolites) dominates and little or no cell damage occurs because the metabolic products are relatively non-toxic and are readily excreted from the body. But as soon as the capacity of the cell to detoxify is overcome because of excessive concentrations of bromobenzene, the dangerous Path A begins to operate and cell damage ensues.

Toxic metabolites are common and toxicologists are learning that many if not most types of toxic, and even carcinogenic, damage are actually brought about by metabolites. Additional examples of this phenomenon surface in later chapters.

As with absorption and distribution, the nature and rate of metabolic transformations vary among individuals and different animal species. Metabolism differences can be extreme, and may be the most important factor accounting for differences in response to chemical toxicity among animal species and individuals within a species. The more understanding toxicologists acquire of metabolism, the more they shall understand the range of responses exhibited by different species and individuals, and the better they shall be able to evaluate toxic risks to humans.

Excretion

Most chemicals and their metabolites eventually depart the body; they are excreted, as shown in Figure 1. The speed at which they leave varies greatly among chemicals, from a few minutes to many years. Because a chemical leaves the body quickly does not mean it is not toxic; damage to components of cells from certain chemicals can occur very quickly. Likewise, because a chemical is poorly excreted does not mean it is highly toxic, although it is true that long residence times can increase the chance of an adverse event occurring.

As blood moves through the kidneys, chemicals and their metabolites can be filtered out or otherwise lost from the blood by a set of extraordinary physiological mechanisms that release them into urine. Urinary excretion is probably the pathway out of the body for most chemicals.

Gases and highly volatile chemicals can move out of the blood into the lungs, and be exhaled. Carbon dioxide, for example, is a metabolic product of many chemicals and also derives from the metabolism of essential molecules; it is excreted from the body through the lungs.

Chemicals may also be excreted in bile. Bile is a fluid normally excreted by the liver. It is composed of some degradation products of normal metabolism, and is excreted out of the liver and into the GI tract. Some chemicals move into bile, out into the GI tract, and get excreted in feces (along with chemicals that are not absorbed from the GI tract, as discussed in the Absorption section earlier).

Interestingly, some metabolites undergoing biliary excretion are reabsorbed, usually after undergoing further metabolic change brought about by enzymes associated with microorganisms normally found in the intestines. There are notable examples of this phenomenon, and it can be important as a factor in toxicity production, but its discussion is beyond the scope of this book.

Some minor routes of excretion exist: sweat, saliva, semen, milk. While these routes out of body do not count for much as excretory processes, excretion of some chemicals into milk can be important because it constitutes an *exposure pathway* for infants, if the milk is from their mothers, and for many people if it is from dairy cattle. PCBs and other fat soluble chemicals follow this pathway out of the body, dissolved in the fatty portion of the milk. Excretion of chemicals through milk is common enough to prompt considerable attention from toxicologists.

Uses of ADME data

Toxicologists generally believe that comprehensive ADME data can provide extraordinarily useful information to assist judging the risks posed by chemical agents. The reasons for this belief are complex, and the best we shall be able to do in this book is to illustrate some of the uses of these data when we discuss specific toxic agents. Up until now our purpose has been simply to create some understanding of how chemicals get into, move about, and get eliminated from the body, how they undergo chemical changes, and to suggest why these processes are important determinants of toxicity.

A very important issue has not been mentioned throughout this discussion: it is generally not possible to acquire comprehensive

ADME data in human beings! To study ADME systematically, and to develop reliable data, requires studies of a type that simply can not be ethically performed in human beings, at least with chemicals of more than a very low toxic potential or clear benefit to the persons involved, as might be the case with certain drugs. It is possible to acquire a little ADME data by, for example, conducting careful chemical analysis on exhaled air or urine of people (sampling of other fluids or tissues, except perhaps for blood, obviously can not be performed) known to be exposed to certain chemicals – say in the workplace. Such analyses may reveal the amounts and chemical identities of excreted metabolites, but little else. Such information may nevertheless be useful, because toxicologists can compare the pattern of urinary metabolites observed in humans and in experimental animals. If the patterns are similar, this tends to support the proposition that toxic effects observed in the species of experimental animals may be relevant to humans, whereas a substantially different pattern may suggest the opposite. In general, human ADME data, except for some pharmaceutical agents extensively studied in clinical settings, are quite limited for environmental chemicals.

It is, of course, possible to obtain comprehensive ADME data through studies in experimental animals, but even this is now available for only a relatively few chemicals – drugs have been the best studied – probably because their utility in evaluating the adverse human health effects of chemicals is less well-established than data obtained from other types of experimental studies. Still, toxicologists value greatly ADME data when they are available and are continually seeking ways to use them to improve understanding of human chemical risks. Uses of ADME data in evaluating cancer risks to humans are reviewed in Chapters 8 and 10.

It is now time to pursue a discussion of the types of injury chemicals and their metabolites can produce as they move about the body, and of how the extent of injury is influenced by the circumstances of exposure.

4

Toxicity and Toxic Risk

Whence toxicology

It is difficult to doubt that the earliest human beings, and perhaps even some of their evolutionary predecessors, were aware of poisons in their environments. For as long as human beings have walked the earth they have been stung or bitten by poisonous insects and animals. In their search for nourishing foods, mostly through trial and error, some early members of the species no doubt were sickened by or even succumbed to the consumption of the many plants that contain highly toxic constituents. Somewhere in prerecorded time, human beings also learned that certain plants could alleviate pain or remedy certain afflictions; learning about these plants probably also taught them a great deal about unpleasant side effects.

Those early metallurgists who were clever enough to learn how to transform crude ores to shiny metals were probably also observant enough to discover that some of the materials being worked with could harm them. Some of the earliest written accounts of man on earth provide evidence that the ancient Greeks and Romans were well aware of the poisonous properties of certain plants and metals. The case of the poisoning of Socrates with hemlock is only the most famous of the early references to the deliberate use of certain plants for suicidal or homicidal purposes.

The science of toxicology, which we define as the study of the adverse effects of chemicals on health and of the conditions under which those effects occur, has begun to take on a well-defined shape only in the past four to five decades. The science is still struggling for a clear identity, but it has begun to find one. One of the reasons for its uncertain status is that toxicology has had to borrow principles and methodologies from several more basic sciences and from certain medical specialties. Several disparate historical strands of the study of poisons appear finally to be intertwining to create the science of modern toxicology.

One strand, and probably the earliest to appear in a systematic form, was the study of antidotes. The modern strand of this field of study is called clinical toxicology. Clinical toxicologists are typically physicians who treat individuals who have suffered deliberate or accidental poisoning. Poisoning as a political act was extremely common through at the least the Renaissance, and its use for homicidal purposes continues on a not insignificant scale to this day. Some of the earliest physicians, including Dioscorides, a Greek who served Nero, and the great Galen himself, were engaged to identify ways to reverse poisonings or to limit the damage they might cause. The Jewish philosopher Maimonides published *Poisons and Their Antidotes* in Arabic, in the year 1312; this text synthesized all knowledge available at the time and served as a guide to physicians for several centuries. The Spanish physician Mattieu Joseph Bonaventura Orfila published in 1814–15 a comprehensive work entitled *A General System of Toxicology or, a Treatise on Poisons, Found in the Mineral, Vegetable, and Animal Kingdoms, Considered in Their Relations with Physiology, Pathology, and Medical Jurisprudence*. Orfila's organization of the topic is considered a seminal event in the history of toxicology.

A second major strand, closely linked to the first, is in the domain of what is today called pharmacology: the study of drugs and of their beneficial and adverse effects. This strand is also a very ancient one. Pharmacologists, many of whom were also botanists, at first collected plants and made catalogues of their beneficial and harmful effects. Some of these works are magnificent compilations of highly detailed information and have proved to be of enormous benefit to mankind. Major advances in pharmacology were brought about by the work of Paracelsus (*c.*1493–1541), a Swiss physician and alchemist who promoted theories of disease that were an odd mix of scientifically advanced notions and fanciful superstitions. Among toxicologists and pharmacologists he is noted for his recognition that 'All substances are poisons; there is none which is not a poison. The right dose differentiates a poison and a remedy.' This remark adorns the frontispiece of almost every toxicology text. It was only in the late 19th and early 20th centuries, however, that pharmacologists began to acquire some understanding of the nature of the specific chemical constituents of plants that were biologically active, in both beneficial and harmful ways. Once chemical science had undergone its revolution, it was possible for pharmacologists to begin to understand how molecular structure influenced biological action.

The tools for the systematic study of the behavior of these drug molecules in biological systems also came under rapid development during this same period, as the science of experimental medicine began to blossom. Pharmacologists drew upon the advances in medicine, biochemistry, and physiology resulting from the work of the great French physiologist Claude Bernard (1813–78) and began to create our modern understanding of drug action and drug toxicity. Some of the principal experimental tools used by modern toxicologists were first developed by pharmacologists. The methods used to collect ADME data, for example, were brought to perfection by scientists studying drug behavior. Some pharmacologists even contend that toxicology is merely a branch of pharmacology.

A third historical strand that has helped to create modern toxicology consists of the labors of occupational physicians. Some of the earliest treatises on toxicology were written by physicians who had observed or collected information on the hazards of various jobs. The man some have called the father of the field of occupational medicine was Bernardino Ramazzini, an Italian physician whose text *De Moribus Artificum Diatriba* (1700) contributed enormously to our understanding of how occupational exposures to metals such as lead and mercury could be harmful to workers. Ramazzini also recognized that it was important to consider the possibility that some poisons could slowly build up in the body and that their adverse effects do not make themselves apparent for a long time after exposure begins.

Sir Percival Pott published in 1775 the first record of occupationally-related human cancers; this London physician recognized the link between cancer of the scrotum and the occupation of chimney sweep. More of Sir Percy's work will be described in Chapter 7.

Occupational physicians, Pott among them, contributed greatly to the development of the modern science of epidemiology – the systematic study of how diseases are distributed in human populations and of the factors that cause or contribute to them. Epidemiology is an important modern science, and its study, as we shall see, can provide toxicologists with highly useful data about the toxic effects of environmental chemicals.

Occupational toxicology and industrial hygiene took a great leap forward in the early part of this century when Alice Hamilton (1869–1970), a physician from Fort Wayne, undertook with enormous energy and unyielding commitment an effort to call national attention to the plight of workers exposed to hazardous substances in mines, mills, and smelting plants throughout the country. Her work, which has perhaps been insufficiently acclaimed, led to renewed interest in occupational medicine and was also instrumental to the introduction of worker's compensation laws. Dr Hamilton became in 1919 the first woman to receive a faculty appointment at Harvard University.

Observations from the field of occupational medicine also created interest among biologists in the field of experimental carcinogenesis – the laboratory study of cancer development – that is the topic of Chapters 7 and 8.

Studies in the science of nutrition make up another strand leading to modern toxicology. The experimental study of nutrition, another offspring of the explosion in experimental medicine that took place following the work of Claude Bernard led, among other things, to increased appreciation of the proper use of experimental animals to understand the biological behavior of nutrients and other chemicals. The pioneering work on vitamins of Philip B. Hawk in the World War I era led to the development of experimental animal models for studying the beneficial and harmful effects of chemicals. One of Hawk's students, Bernard Oser, who is today still active in the field, contributed enormously to the perfection of the experimental animal model, particularly in connection with the study of the toxicology of foods and food ingredients. Work in experimental nutrition also gave toxicologists a sense of the importance of individual and species variability in response to exogenous chemicals (i.e., chemicals entering the body from an outside source).

Modern toxicologists have also drawn upon the work of radiation biologists – scientists who study the biological effects of various forms of radiation. The development of radiation biology spawned important work on the genetic components of cells and the ways in which they might be damaged by environmental agents. It also provided insight into some of the biological processes involved in the development of cancers from damaged, or mutated, cells. The Manhattan Project itself created a need to understand the toxic properties of the myriad of chemicals that were then being prepared and handled in unprecedented amounts. Stafford Warren, head of the Department of Radiology at the University of Rochester, established a major research program on the toxicology of inhaled materials, including radioactive substances; Warren's group included Herbert Stokinger, Harold Hodge, and several other scientists who went on to become luminaries in the field.

For the past 100 years, and particularly since about 1935, toxicologists have been activated by developments in synthetic chemistry. As the chemical industry began to spew forth hundreds and thousands of new products, pressures were created for the development of information about their possible harmful effects. The first federal law that gave notice of a major social concern about poisonous products was the Pure Food and Drug Act, passed by Congress in 1906 and signed into law by Theodore Roosevelt. Much of the impetus for the law came from the work of Harvey Wiley, chief chemist of the Department of Agriculture, and his so-called 'Poison Squad.' Wiley and his team of chemists not infrequently dosed themselves with suspect chemicals to test for their deleterious effects.

The systematic study of toxic effects in laboratory animals began in the 1920s, in response to concerns about the unwanted side effects of food additives, drugs, and pesticides (DDT and related pesticides became available in this era). Concerns uncovered during the 1930s and 1940s about occupational cancers and other chronic diseases resulting from chemical exposures prompted increased activity among toxicologists. The modern version of the Food, Drug, and Cosmetic Act was enacted by Congress in 1938 in response to a tragic episode in which more than 100 people died from acute kidney failure after ingesting certain samples of the antibiotic sulfanilamide ('Elixir of Sulfanilamide') that had been improperly prepared in a diethylene glycol solution. Diethylene glycol is obviously better suited for its use as antifreeze. This law was only the first of many that have contributed to the creation of the modern science of toxicology. The Elixir of Sulfanilamide tragedy had a beneficial side effect, in that it prompted some of the earliest investigations into underlying mechanisms of toxicity by Eugene Geiling at the University of Chicago. The scientists Geiling gathered to work on the diethylene glycol problem and on other emerging problems in toxicology were to become leaders of the field during the following three decades.

Sporadically during the 1940s and 1950s the public was presented with a series of seemingly unconnected announcements about poisonous pesticides in their foods (most infamous of which was the great cranberry scare of 1959, in which federal officials announced, just before Thanksgiving, that it would be 'prudent' to avoid consuming these berries because they were contaminated with a carcinogenic herbicide), food additives of dubious safety, chemical disasters in the workplace, and air pollution episodes that claimed thousands of victims in urban centers throughout the world. In 1962 Rachel Carson, a biologist from Silver Spring, Maryland, drew together these various environmental horror stories in her book, *Silent Spring*. Carson wrote:

> For the first time in the history of the world, every human being is now subjected to contact with dangerous chemicals, from the moment of conception until death. In the less than two decades of their use, the synthetic pesticides have been so thoroughly distributed throughout the animate and inanimate world that they occur virtually everywhere.

and

> Human exposures to cancer-producing chemicals (including pesticides) are uncontrolled and they are multiple. ... It is quite possible that no one of these exposures alone would be sufficient to precipitate malignancy – yet any single supposedly 'safe dose' may be enough to tip the scales that are already loaded with other 'safe doses'.

Silent Spring was immensely popular and influential. Carson's work almost single-handedly created modern society's fears about synthetic chemicals in the environment and, among other things, fostered renewed interest in the science of toxicology. It also helped pave the way for the introduction of several major federal environmental laws in the late 1960s and early 1970s, and for the creation of the EPA in 1970.

Even this fragmentary and superficial history should provide some sense of why the science of toxicology has had identity problems. Its creation has been in part forced by social pressures, such as those created by the publication of *Silent Spring*, that are now at their historically most intense, and that have moved the scientific community to attempt to acquire knowledge about the harmful effects of chemicals in the environment. Most of these pressures concern man-made chemicals although, as I hope to show later in the book, there is not much basis for distinguishing man-made chemicals from naturally occurring chemicals as agents of harm.

Beginning in the 1930s individuals from a wide variety of scientific and medical disciplines, and working in various government, industry, and academic laboratories, began drawing upon the accumulated experience of the many fields of study that had been devoted to understanding the behavior of chemical substances, including some having radioactive properties, when they came into contact with living systems. Other than the physicians who had specialized in clinical toxicology and medical forensics, these individuals were not 'toxicologists' – the discipline did not exist. They were pharmacologists and biochemists, physiologists, general biologists, epidemiologists, experimental nutritionists, pathologists, and scientists involved in experimental cancer studies. Scientists from the basic disciplines of chemistry and physics, and statisticians specializing in biological phenomena, also made contributions.

Out of this collective effort, a coherent scientific enterprise began to take shape. The first major professional group under which these investigators collected and published their work was The Society of Toxicology, founded as recently as 1961. A few graduate schools and schools of public health began offering advanced degrees in toxicology in the 1960s, and toxicology courses began to be included in other, related graduate curricula. Professional journals began to multiply and several became internationally prominent. The National Academy of Sciences began to deal with toxicology issues in the 1940s, and they have since become a major feature of Academy efforts. These trends, and several others to be mentioned later in the book, have led to the common acceptance of a set of definitions, principles, and methodologies that guide the discipline. While considerable debate on certain matters exists, consensus or near-consensus exists on many others. The principal purpose of the remainder of this chapter is to set forth some of the important definitions, principles, and methodologies that have emerged and become firmly established over the past few decades. This discussion will set the stage for the later chapters, in which specific types of toxicity and toxic agents are examined.

Activities of toxicologists

In brief, toxicologists are involved in three types of activities:

(A) The study of the types of adverse health effects produced by chemicals under various conditions of exposure.
(B) The study of the underlying biological events, including but not limited to ADME studies, by which chemicals create adverse health effects.
(C) The evaluation of information on the toxic properties of chemicals, the conditions under which these properties manifest themselves, and the underlying biological processes leading to toxicity, to assess the likelihood that those chemicals might produce their adverse effects in human populations that are or could be exposed to them.

Toxicologists engaged in activities (A) and (B) are typically laboratory scientists. Even today many involved in these efforts were not trained as toxicologists, but come to them from a variety of disciplines. Moreover, some aspects of these activities require a degree of specialization, in disciplines such as pathology and statistics, that most toxicologists do not have.

Toxicologists engaged in activity (C) do so outside the laboratory. They may undertake such activities as members of various expert committees, as employees of regulatory agencies, or as scientists in corporations who are responsible for giving advice to management on matters of chemical risks. Activity (C) is called risk assessment; it is a difficult, controversial, and unsettled area to which Chapter 10 is devoted. Toxicologists engaged in risk assessment are typically aided by epidemiologists, statisticians, experts in human exposure analysis, and other toxicologists whose principal occupations are activities (A) and (B).

It should be clear, even from this somewhat oversimplified picture of what toxicologists do, that they do not do it all alone; toxicology is a discipline, but a thorough evaluation of the risks of chemical agents requires a multidisciplinary effort, and even today it is not possible clearly to define the boundaries between toxicology and the several disciplines it draws upon. To make things simple, however, we shall from now on refer to toxicologists and the discipline of toxicology as if we were confident we understood what these terms mean.

Some important terms and principles

All chemicals, natural and synthetic, are toxic – that is, they produce adverse health effects – under some conditions of exposure. It is incorrect (but I'm afraid very common) to refer to some chemicals as toxic and others as non-toxic. If this book teaches any lesson, it is that this notion is not correct.

Chemicals do, however, differ greatly in their capacity to produce toxicity. The *conditions of exposure* under which toxic effects are produced – the size of the dose and the duration of dosing needed – vary greatly among chemicals. Moreover, the nature and severity of the toxic effects produced are also highly varied, and are different not only for different chemicals, but also for a single chemical as the conditions of exposure to it change.

A commonly used scheme for categorizing toxicity is based on exposure duration. Toxicologists generally seek to understand the effects of *acute*, *chronic*, and *subchronic* exposures. They attempt to learn for each of these three exposure categories the types of adverse effects a chemical produces, the minimum dose at which these effects are observable, and something about how these adverse effects change as the dose is increased.

Acute exposure involves a single dose. Toxicologists frequently refer to the adverse consequence of an acute exposure as an 'acute effect'. This usage is not incorrect, but, as will be seen in a moment, it can be misleading when similarly applied to chronic exposures.

In studying the acute toxicity of a chemical our interest is in understanding the dose that will lead to some harmful response and also to the most harmful one of all, death. The next chapter ('Fast Poisons') is devoted to acute poisoning. We shall see that some chemicals produce toxicity and death after a single exposure at extremely low doses, while others do so only at doses that are so high they are nearly impossible to get into the body. Most chemicals fall between these extremes. The notion that the world consists of two neatly separated categories of chemicals, the toxic and the non-toxic, derives largely from the notion that for many chemicals (the 'non-toxic' ones) extremely high and unlikely doses are needed to produce acute toxicity. This toxic/non-toxic dichotomy, while as a practical matter useful for separating the substances we should be concerned about for their acute effects from those we need not worry about, creates a highly misleading impression about the nature of chemical risks.

Chronic exposure generally refers to dosing over a whole lifetime, or something very close to it. Subchronic is less well-defined, but obviously refers to repeated exposures for some fraction of a lifetime. In animal toxicity studies involving rodents, chronic exposures generally refer to daily doses over about a two-year period, and subchronic generally refers to daily doses over 90 days. Again, for both these exposure durations, toxicologists are seeking to learn the specific types of adverse effects associated with specific doses of the chemical under study. Some dosing regimens do not fall usefully into the chronic or subchronic categories, and some of these will be encountered in later discussions of the effects of chemicals on the reproductive process. The toxic effects of subchronic and chronic exposures are covered in Chapter 6 ('Slow Poisons') and Chapter 7 ('Carcinogens').

Care must be taken to distinguish subchronic or chronic exposures from subchronic or chronic effects. By the latter, toxicologists generally refer to some adverse effect that does not appear immediately after exposure begins but only after a delay; sometimes the effect may not be observed until near the end of a lifetime, even when exposure begins early in life (cancers, for example, are generally in this category of chronic effects). But the production of chronic effects may or may not require chronic exposure. For some chemicals acute or subchronic exposures may be all that is needed to produce a chronic toxicity; the effect is a delayed one. For others chronic exposure may be required to create chronic toxicity. Toxicologists are not always careful to distinguish subchronic and chronic exposures and effects.[5] In this book we shall refer to exposures as subchronic or chronic, and talk about toxic effects as immediate (quickly following exposure) or delayed.

Toxicologists refer to *targets* of toxicity. Some chemicals damage the liver, others the kidney, and some damage both organs. Some adversely affect the nervous system or the reproductive system or the immune system or the cardiovascular system. The brain, the lungs, elements of the blood, the blood vessels, the spleen, the stomach and intestines, the bladder, the skin, the eye – all can be damaged by chemical agents. Toxicity may be exerted by some chemicals on the developing embryo and fetus. It is convenient to categorize chemicals by the organ or system of the body that is the target for their toxicity, so we refer to liver toxicants, nervous system toxicants, dermal toxicants, and so on. Some chemicals will fall into only one of these categories, but most fall into several. Moreover, as exposure conditions change, so may targets.

This type of categorization, while convenient, might be misleading. It perhaps suggests that all chemicals having a common target produce the same type of toxic effect on that target. This is not the case, and we shall reveal several examples in the next two chapters.

Chemicals causing certain adverse effects are singled out for special treatment. Those capable of producing excess tumors in any of many possible sites of the body are classified as *carcinogens*, and not according to the target at which they act (although they may be subcategorized as lung carcinogens, liver carcinogens, etc.). Chemicals causing birth defects of many different types are classified as *teratogens* (from the Greek 'teras,' meaning 'monster'). Some chemicals alter behavior in undesirable ways, and so are classified as behavioral toxicants. These are a few of the special categories toxicologists have come to rely upon as they go about organizing their knowledge.

---

[5] Physicians, of course, refer to diseases as chronic if they persist a long time in the patient, or if they have been a long time in development. This is a perfectly appropriate usage.

Some toxic effects are reversible. Everyone has been exposed to some agent, household ammonia for example, that produces irritation to the skin or eyes. Exposure ends and, sometimes perhaps with a delay, the irritation ends. Some readers have no doubt been poisoned on occasion by the ingestion of too much alcohol. The effects here also reverse. The time necessary for reversal can vary greatly depending upon the severity of the intoxication and certain physiological features of the person intoxicated. But most people also realize that chronic alcohol abuse can lead to a serious liver disorder, cirrhosis, which may not reverse even if alcohol intake ceases. This type of effect is irreversible or only very slowly reversible. It is important in making a toxicological evaluation to understand whether effects are reversible or irreversible, because one is obviously much more serious than the other.

Risk

In the final analysis we are interested not in toxicity, but rather in *risk*. *By risk is meant the likelihood, or probability, that the toxic properties of a chemical will be produced in populations of individuals under their actual conditions of exposure.* To evaluate the risk of toxicity occurring for a specific chemical at least three types of information are required:

(1) The types of toxicity the chemical can produce (its targets and the forms of injury they incur).
(2) The conditions of exposure (dose and duration) under which the chemical's toxicity can be produced.
(3) The conditions (dose, timing and duration) under which the population of people whose risk is being evaluated is or could be exposed to the chemical.

It is not sufficient to understand any one or two of these; no useful statement about risk can be made unless all three are understood. It may take the next several chapters to create a thorough understanding of these matters, but they are the heart of the lesson of this book.

5

Fast Poisons

Rating acute toxicity

Botulinum toxins are a collection of protein molecules that are exquisitely poisonous to the nervous system. These toxins[6] are metabolic products of a common soil bacterium, *Clostridium botulinum*, which is frequently found on raw agricultural products. Fortunately, the bacterium produces its deadly toxins only under certain rather restricted conditions, and if foods are properly processed so that these conditions are not created, the toxins can be avoided. Food processors have to be extremely careful with certain categories of food – canned foods having low acidity, for example – because the slightest contamination can be deadly. Most reported cases of botulism have involved vegetables improperly canned in the home.

Botulinum toxins can be *lethal* at a single (acute) dose in the range of 0.000 01 mg/kg b.w.! This amount of toxin is not visible to the naked eye. The initial symptoms of botulism typically appear 12–36 hours after exposure and include nausea, vomiting, and diarrhea. Symptoms indicating an attack on the nervous system include blurred vision, weakness of facial muscles, and difficulty with speech. If the dose is sufficient (and a very, very small dose can be), the toxicity progresses to paralysis of the muscles controlling breathing – the diaphragm. Death from botulism thus comes about because of respiratory failure. Botulinum toxins are regarded as the most acutely toxic of all poisons.

Sucrose, which we all know as table sugar, can also be acutely toxic.

---

[6] The name 'toxin' is correctly applied to naturally occurring protein molecules that produce serious toxicity. There has been some tendency to broaden use of this term to include other categories of toxic agents. We shall adhere to the proper usage in this book.

Table 1. *Conventional rating scheme for lethal doses in humans*

Toxicity rating	Probable lethal oral dose for humans	
	Dose (mg/kg b.w.)	For average adult
1 Practically non-toxic	more than 15 000	More than 1 quart
2 Slightly toxic	5000–15 000	1 pint–1 quart
3 Moderately toxic	500–5000	1 ounce–1 pint
4 Very toxic	50–500	1 teaspoon–1 ounce
5 Extremely toxic	5–50	7 drops–1 teaspoon
6 Supertoxic	less than 5	less than 7 drops

I can't locate any evidence of humans being killed with a dose of table sugar, but toxicologists can force enough into rats to cause death. A lethal dose of sucrose in rats is in the range of 20 000 mg/kg b.w. That's about as 'non-toxic' as chemicals get to be. If humans are equally sensitive, one would have to eat more than 3 pounds of sugar at one time for it to be lethal.

The two substances – sucrose and botulinum toxins – differ in lethality by about 10 billion times! The acute lethal doses of most chemicals fall into a much narrower range, but there are many substances near the two extremes of this distribution of lethal doses.

Clinical toxicologists have found it convenient to rate chemicals according to their potential to produce death after a single dose. The conventional rating scheme is as shown in Table 1.

Clearly, if a person has to ingest a pint or more of a chemical before his life is seriously threatened, this chemical is not a likely candidate for use in homicide or suicide, and is highly unlikely to be accidentally ingested in dangerous amounts. Chemicals rated in categories 5 or 6, however, need to be extremely carefully controlled.

Keep in mind that the rating chart presented above concerns only *acute, lethal* doses, received by the oral route. It provides only a highly limited picture of the toxic properties of chemical agents, and should never be used as the sole basis for categorizing chemicals. Some chemicals that are 'supertoxic' by the above rating have no known detrimental effects when they are administered at sublethal doses over long periods of time, while others in the same category and in lower categories do produce serious forms of toxicity after repeated dosing.

Table 2. *Some supertoxic chemicals**

Chemical	Source	Principal toxicity target
Botulinum toxin	Bacterium	Nervous system
Tetrodotoxin	Puffer fish (fugu)	Nervous system
Crotalus venom	Rattle snake	Blood/nervous system
Naja naja	Cobra	Nervous system/heart
Batrachotoxin	South American frog	Cardiovascular system
Stingray venom	Stingray	Nervous system
Widow spider venom	Black widow	Nervous system
Strychnine	Nux vomica**	Nervous system
Sodium fluoroacetate	Synthetic chemical	Heart and nervous system (inhibits energy metabolism)
Nicotine	Tobacco plant	Nervous system
Phosgene	Synthetic chemical	Respiratory system

* Some of these chemicals, particularly the venoms, are protein or protein-like compounds that are deactivated in the gastrointestinal tract; they are poisonous only when injected directly into the blood stream.
** The seed of the fruit of an East Indian tree used as a source of strychnine.

Some highly toxic chemicals

A few 'extremely toxic' and 'supertoxic' chemicals are listed in Table 2, along with their environmental sources and toxicity targets. Some of these are toxins found in the venoms of poisonous snakes or in the tissues of certain species of animals. Dr. Findlay Russell, who has made enormous contributions to our understanding of the nature of animal toxins, their modes of biological action, and the procedures for treating people who have been envenomed or poisoned, estimates that there are about 1200 known species of poisonous or venomous marine animals, 'countless' numbers of venomous arthropods (spiders), and about 375 species of dangerous snakes (out of a total of about 3500 species).

We have been speaking of both 'venomous and poisonous' animals, and there is a distinction between the two. A venomous animal is one that, like a snake, has a mechanism for delivering its toxins to a victim, usually during biting or stinging. A poisonous animal is one that contains toxins in its tissues, but cannot deliver them; the victim is poisoned by ingesting the toxin-containing tissue.

An interesting and important example of an animal poison is paralytic shellfish poison (PSP). This chemical (perhaps it is a group of chemicals), which is also known as saxitoxin and by several other names as well, is found in certain shellfish. But it is not produced by shellfish; it is rather a metabolic product of certain marine microorganisms (protista). These microorganisms are ingested by the shellfish as food, and their poison can remain behind in the shellfish's tissue. PSP is not a protein, but a highly complex organic chemical of most unusual molecular structure.

Shellfish accumulate dangerous levels of PSP only under certain conditions. Typically, this occurs when the microorganisms undergo periods of very rapid growth, resulting from the simultaneous occurrence of several favorable environmental conditions. This growth, or 'bloom,' frequently imparts a red color to the affected area of the ocean, and is referred to as a Red Tide. Shellfish growing in a Red Tide area can accumulate lethal amounts of PSP.

Red Tides (and some with other colors as well) occur with some regularity in certain coastal waters of New England, Alaska, California, and several other areas. Public health officials typically seek to quarantine affected areas to prevent harvesting of shellfish. In some areas of the Gulf of Alaska, large reservoirs of shellfish cannot be used as food because of a persistent PSP problem.

PSP, like botulinum toxin, is a neurotoxic substance (it affects the nervous system) and can also affect certain muscles, including the heart. Some poisoned humans who recovered from the effects of PSP have described the early stages of intoxication as not at all unpleasant: a tingling sensation in the lips and face and a feeling of calm. Those who die from PSP ingestion do so because of respiratory failure.

The first successful method for measuring the amount of PSP in shellfish was published in 1932. The procedure used was a *bioassay*. A bioassay is a measurement procedure in which some living organism is used to detect the presence of a biologically active agent. The toxicity of chemicals is measured using various types of bioassays, so we shall have much to say about them.

The bioassay for PSP was simple. Extracts from shellfish suspected of contamination were fed to mice. The poison was measured in 'mouse units.' A mouse unit was the amount of toxin that would kill a 20 gram mouse in 15 minutes. Crude, but nevertheless effective at telling public health officials when shellfish were too toxic to eat.

The plant kingdom is another source of some unusually toxic chemicals. A few examples are presented in Table 3, along with a description of some of their biological effects.

Infants and preschoolers are the most frequent victims of plant toxins. Their natural curiosity leads them to put all sorts of non-food

Table 3. *Poisonous properties of some common plants*
(The specific chemicals involved are in many cases not known)

Plant	Effects
Water hemlock	Convulsions
Jimson weed	Many, including delirium, blurred vision, dry mouth, elevated body temperature
Foxglove, lily-of-the-valley, oleander	Digitalis poisoning – cardiovascular disturbances
Dumbcane (dieffenbachia)	Irritation of oral cavity
Jonquil, daffodil	Vomiting
Pokeweed	Gastritis, vomiting, diarrhea
Castor bean	Diarrhea, loss of intestinal function, death
Poison ivy, poison oak	Delayed contact sensitivity (allergic dermatitis)
Potatoes, other solanaceous plants*	Gastric distress, headache, nausea, vomiting, diarrhea

*The solanaceae include many species of wild and cultivated plants, the latter including potatoes, tomatoes, and eggplants. All these plants contain certain natural toxicants called solanine alkaloids. The levels found in the varieties used for food are below the toxic level, although not always greatly so. Potatoes exposed to too much light can begin to grow and to produce excessive amounts; the development of green coloring in such potatoes (chlorophyll) indicates this growth. Storing potatoes under light, particularly fluorescent light, is to be discouraged.

items into their mouths, and berries, flowers, and leaves from house and yard plants are often attractive alternatives to spinach. The number of deaths from consumption of poisonous plants is not great, but the number of near-deaths is; about 10 per cent of inquiries to Poison Control Centers concern ingestion of house, yard, and wild plants, including mushrooms. Among the house plants dumbcane (species of dieffenbachia) and philodendrons are prominent, and a fair number of poisonings arise from jade, wandering Jew, poinsettia, schifflera, honeysuckle, and holly.

Children are also especially vulnerable for a reason touched upon in Chapter 2. Consider the family of mushroom toxins known as amatoxins. Almost all mushroom-related deaths in North America are caused by these toxins, which are metabolic products of *Amanita phalloides*. These toxins are a bit unusual because symptoms appear only after 12 hours following ingestion; they include vomiting, diarrhea, and very intense abdominal pain. Ultimately the toxins cause liver injury that can be serious enough to cause death.

The lethal dose of amatoxins is in the range of 1 mg/kg b.w. For an adult weighing 70 kg, a total of about 70 mg needs to be consumed to cause death (1 mg/kg × 70 kg). For a one-year old child weighing 10 kg, only about 10 mg needs to be ingested to create a life-threatening intake of the toxins. Small body size is highly disadvantageous.

There are also some synthetic chemicals that display supertoxic properties. Among these are certain so-called organophosphorus insecticides that inhibit the normal biological action of a certain enzyme that is a critical participant in nerve impulse transmission. The most highly toxic members of this class of compounds are not suitable for use as insecticides – they can seriously threaten humans who are handling them – but are included in the catalogue of chemical warfare agents, the so-called 'nerve gases.'

A curious and environmentally important byproduct of industrial society is a chemical called 2,3,7,8-tetrachlorodibenzo-*p*-dioxin, which is usually simply called 'dioxin.' Its chemical structure is as shown. More will be said about dioxin in the chapter on carcinogens, but here we are interested only in its acute toxicity. Dioxin is, indeed, highly toxic and can cause deaths in female guinea pigs at a single dose of 0.0006 mg/kg b.w. What is curious about dioxin is that it does not kill hamsters unless the dose is about 5000 times greater! Now it is not uncommon for different species to react differently to the same chemical, but this type of extreme difference in lethal dose is odd. No one understands why dioxin behaves this way (or why different species react so differently when exposed to dioxin), but, fortunately, human beings appear to fall near the resistant end of the range of sensitivities.

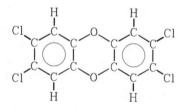

2,3,7,8−tetrachlorodibenzo−*p*−dioxin

There are hundreds of extraordinarily interesting tales about highly toxic chemicals, but to say more on this topic gets us too far adrift from our main course. A couple of important points have been made here. First, virtually all chemicals can cause a deadly response after a single dose, but enormous differences exist among chemicals in the capacity to kill. Categorizations of chemicals as 'extremely toxic', 'practically non-toxic', and so on, generally refer only to this limited aspect of their toxic potential, and should not be used as general indicators of the full range of toxic effects a chemical may cause.

That nature is not benign, indeed that it can be extremely deadly, is another lesson to take away from this section. While chemists have been successful in creating some highly lethal agents, none of these quite matches the likes of botulinum toxin, cobra venom or the poison in puffer fish (tetrodotoxin).

Although man-made chemicals, or the incidental byproducts of industrial society do not create quite the risk of acute lethality that some natural poisons do, there are some that can cause considerable non-lethal toxicity after a single exposure. Most of these acute exposures are created by industrial or transportation accidents, which are not infrequent. To minimize the damage accidental releases such as these might cause, it has become important to use the toxicity rating classification discussed earlier and to label industrial chemicals accordingly to ensure appropriate care is taken in handling, storing, and transporting them. Of course, there have been industrial accidents involving releases sufficiently large to cause deaths, sometimes to workers, sometimes to nearby residents or passersby. The worst example of this type of event took place during the night of December 3, 1984, at Bhopal, India. Approximately 40 tons of methyl isocyanate (a very simple organic chemical used in the synthesis of an important pesticide and having the structure shown) were released into the atmosphere, killing more than 2000 people and injuring many more.

$$H-\underset{\underset{H}{|}}{\overset{\overset{H}{|}}{C}}-N=C=O$$

Methyl isocyanate

Acute exposures leading to non-lethal effects

Threats to health other than death are obviously not to be ignored. Some threats are more worrisome than others. Depending upon both the chemical and the size of the dose, acute exposures can lead to everything from minor and completely reversible irritation of the skin or eyes, to serious injuries of many different targets that can threaten life and well-being.

Episodes that lead to acute, non-lethal events are many and varied. They range from accidents in the home with household products to consumption of plant and animal toxins; from industrial and transportation accidents to skin contact with certain plants such as poison ivy; from consumption of too much alcohol to inhaling of certain air pollutants, both outdoors and indoors, that have for some reason accumulated to unusually high levels.

The nature of the adverse effect produced by a single exposure to a chemical will change as the dose changes. Take the example of alcohol. Up to some dose ingestion of alcohol will produce no detectable adverse response. We are talking here, recall, only of the effects of a single exposure, which could take place, of course, over several hours. At some dose, which varies considerably among individuals, the first indications of intoxication appear: incoordination, blurring of the vision, reduced reaction time. At a somewhat higher dose, the effects are more severe, and involve slurred speech, difficulty in walking, and impaired vision. Raise the dose further and the intoxicated person loses all coordination and may even suffer convulsions. At some still higher dose ethyl alcohol will cause coma and death. Except for the last condition, these effects will begin to disappear once the alcohol exposure ceases.

Although the targets will be different and the types of effects will vary, the toxic effects of all chemicals are characterized by what the toxicologist calls a *dose–response* relationship. Increasing dose will increase the severity of the effect; targets may also change as dose increases. This phenomenon, which is so important that a whole chapter is devoted to it, is characteristic not only of acute exposures, but of all others as well.

Effects of exposure routes

Most of the examples presented so far have concerned ingestion – the oral route of exposure. Acute exposures by other routes can also lead to toxicity. Animal or insect venoms may be injected directly into the blood as a result of biting or stinging. Accidental releases to the air of volatile agents can lead to toxicity in the airways or lungs, and there are thousands of examples of agents producing local toxicity on the skin.

Chemicals can be absorbed into the blood after ingestion, inhalation, or skin contact following an acute exposure, and produce toxic effects at internal targets. It is common to observe that the dose required to produce acute toxicity will vary according to exposure route. Thus, for example, benzene can produce lethality when it is inhaled, ingested, or applied to the skin. Because the lethal effects of benzene require the chemical to get to the bloodstream, and thereby to certain critical targets, the route of exposure that creates the highest blood concentration for a given dose of benzene will represent the route of greatest toxicity. This follows directly from the discussion of ADME in Chapter 3. Note also that, for several reasons, some chemicals will produce different types of toxicity by different exposure routes.

Learning about toxicity

It is time to inquire about the methods used to identify the toxic properties of chemicals. So far a few key principles have been introduced and some information on specific substances has been discussed, but little has been said about how these principles and information have been learned. Without some appreciation of the basic methods of toxicology, and what can and cannot be accomplished with them, it will not be possible to gain a solid understanding of the strengths and, more importantly, the limitations in our knowledge of chemical risks.

Because 'fast poisons' are the subject here, we shall begin to deal with the methods of toxicology as they apply to the characterization of acute toxicity. As the discussion proceeds to 'slow poisons' and to carcinogens, more will be said about methodology, because uncovering information about these classes of toxic agents requires methodological approaches unnecessary for or inapplicable to acute toxicity. The strengths and limitations of these methodologies are best understood if they are presented in the context of the specific problem (acute vs. chronic exposure situations, for example) the toxicologist is trying to understand.

Toxic properties are identified in three basic ways: through *case reports*; with the tools of *epidemiology*; and through *laboratory studies*, typically involving animals but also involving microorganisms, cells, and even parts of cells. Laboratory studies are of two types. The first involves what is called the *toxicity test* or *bioassay*, a study in which chemicals are administered in various ways to groups of laboratory animals or other organisms and observations are made on any adverse effects that ensue. The second type of laboratory study involves whole animals or parts of them (collections of cells, for example), and is designed to provide information on ADME and on what is referred to as the *mechanism of toxic action*. By the latter we mean an understanding of the paths taken by the chemical and its metabolites through the body and the various interactions they undergo with targets and portions of targets, particularly with the ingredients of cells, to produce their adverse health effects. Examples of mechanistic studies and how they contribute to an understanding of toxicity and chemical risk will be brought forward later in the book, in connection with discussions of specific agents, particularly carcinogens, and in Chapter 8. For right now our interest in animal data concerns *toxicity testing*.

Case reports are just what they sound like: reports, typically from physicians, regarding individuals who have suffered some adverse health effect or death following exposure to a chemical in their environment. In the typical case exposures have been acute and effects immediate – which means they are easily recognized as having been produced by a specific agent. Case reports usually involve accidental poisonings, drug overdoses, or homicide or suicide attempts. They have been instrumental in providing early signals of the toxic properties of many chemicals, particularly regarding acute toxicity, and occasionally can be valuable sources of information about the effects of chronic or subchronic exposures. Much of the very early information concerning the carcinogenic properties of arsenic, for example, came from physicians who observed unusual skin cancers in some of their patients treated with Fowler's solution, a once widely-used arsenic-containing medicine.

Case reports do not derive from controlled scientific investigations, but rather from careful and sometimes highly sophisticated scientific and medical detective work. Evidence is pieced together from whatever fragments of information are available, and is frequently not definitive. Only in circumstances in which exposure to a particular agent is clear, toxic effects are immediate, and when other possible causes of those toxic effects can be ruled out, is it possible to establish a clear causal relationship between the agent and the effect. Establishing causality when effects are delayed and the exposure situation not clearly understood is generally not possible. It is also difficult in these situations to satisfy the toxicologist's goal of understanding the size of the dose necessary to produce toxicity. Learning just how much exposure took place as the result of an industrial accident is obviously very difficult. So while case reports will continue to provide clues to risks in the environment, they are perhaps the least valuable source of information for identifying toxic properties, except for those involving exposures capable of eliciting immediate and easily identifiable effects.

Epidemiological studies are a far more important source of information on the effects of chemicals in humans. The epidemiologist tries to learn how specific diseases are distributed in various populations of individuals. Attempts are made to discover whether certain groups of people experiencing a common exposure situation (workers engaged in a common activity, for example, or patients taking the same medicine) also experience unusual rates of certain diseases. Epidemiologists may also try to learn whether groups of individuals having a disease in common also shared a specific type of exposure situation. Such studies are not controlled, in the same sense that laboratory studies are controlled. Epidemiologists attempt to take advantage of existing human exposure situations and, by imposing certain restrictions on how data from those situations are to be analyzed, seek to convert them to something approaching a controlled laboratory study. When this works, such studies can provide immensely valuable information about the toxic properties of chemicals in human beings. Particularly in the chapter on carcinogenesis shall we see some striking examples of the epidemiologists' art.

Creating something approximating a controlled study out of a 'natural' exposure situation is, it must be added, fraught with hazards, many of which simply can not be overcome. It is rare that any single epidemiology study provides sufficiently definitive information to allow scientists to conclude that a cause–effect relationship exists between a chemical exposure and a human disease. Instead epidemiologists search for certain patterns. Does there seem to be a consistent association between the occurrence of excess rates of a certain condition (lung cancer, for example) and certain exposures (e.g., to cigarette smoke) in several epidemiology studies involving different populations of people? If a consistent pattern of associations is seen, and other criteria are satisfied, causality can be established with reasonable certainty. (The difference between establishing that two events are 'associated', which is relatively easy, and the difficult task of establishing that one event 'causes' the other, will be more fully discussed in the chapter on carcinogens.)

We would, of course, prefer not to see anything but negative results from epidemiology studies. In an ideal world information on toxic properties would be collected before human exposure is allowed to take place, and that information would be used to place limits on the amount of human exposure that is permissible. If mechanisms existed to enforce those limits, then excess chemical risk would not occur and, it obviously follows, would not be detectable by the epidemiologist (unless, of course, the data or methods for setting limits were in error).

The world is, of course, not ideal, and over the past 100 years human exposures to thousands of commercially-produced chemicals and the by-products of their production have been allowed to occur prior to the development of any toxicity data other than those related to short-term exposures. During the 1950s and 1960s various federal laws were enacted requiring the development of toxicity information prior to the marketing of certain classes of commercial products – food and color additives, pesticides, human drugs. The Toxic Substances Control Act (1976) imposed similar requirements on certain other classes of industrial chemicals. So in the past few decades we have begun to take steps toward that 'ideal' world.

Epidemiology studies are, of course, useful only after human exposure has occurred. For certain classes of toxic agents, carcinogens being the most notable, exposure may have to take place for several decades before the effect, if it exists, is observable – some adverse effects, such as cancers, require many years to develop. The obvious point is that epidemiology studies can not be used to identify toxic properties prior to the introduction of a chemical into commerce. This is one reason toxicologists turn to the laboratory.

Another reason relates to the fact that for methodological reasons, epidemiology studies can not provide telling information in many exposure situations. It is frequently not possible to find a way to study certain situations in meaningful ways. Recognizing the limitations of the epidemiologic method and concerned about ongoing human exposures to the many substances that had been introduced into commerce prior to their having been toxicologically well characterized, scientists began in the late 1920s to develop the laboratory animal as a surrogate for humans.

To learn about toxicity prior to the marketing of 'new' chemicals, and to learn about the toxicity of 'old' chemicals that did not have to pass the pre-market test: these are the two major reasons toxicologists have turned to testing in laboratory animals.

In addition to the fact that animal tests can be applied to chemicals prior to marketing (we haven't said what should be obvious: it is not ethical to 'test' chemicals in humans), such tests hold several advantages over epidemiology studies. First, and most important, is the fact that they can be controlled. We use this term in its scientific sense. Simply put, a toxicity study is controlled if the *only* difference between two groups of experimental animals is exposure to the chemical under study in one group and the absence of such exposure in the other. Only when studies are strictly controlled in this way can it be rigorously established that adverse effects occurring in one group and not in the other are *caused* by the agent under study.

Using laboratory animals also permits the toxicologist to acquire information on all the targets that may be adversely affected by a chemical, something that is not achievable using epidemiological science. Animals can be extensively examined by the toxicologist and the pathologist, whereas the epidemiologist is usually limited to whatever specific diseases are recorded for the population under study and for suitable 'controls'.

Several more advantages of animal studies will be seen in the next chapter, but here, in this general introduction to the subject, their obvious disadvantage should at least be noted – laboratory animals are not *Homo sapiens*. Toxicologists use rats and mice, sometimes dogs, hamsters, guinea pigs, and even monkeys and apes – all mammals having the same basic biological features of humans. Much empirical evidence exists to show that laboratory animals and human beings respond similarly to chemical exposures. Dr. David Rall, former Director of the National Institute of Environmental Health Sciences, and some of his associates put the matter this way:

The adequacy of experimental data for identifying potential human health risks and, in particular, for estimating their probable magnitude has been the subject of scientific question and debate. Laboratory animals can and do differ from humans in a number of respects that may affect responses to hazardous exposures. . . . Nevertheless, experimental evidence to date certainly suggests that there are more similarities between laboratory animals and humans than there are differences. These similarities increase the probability that results observed in a laboratory setting will predict similar results in humans.

Toxicologists do not, however, have convincing evidence that every type of toxic response to a chemical observed in species of laboratory animals will also be expected to occur in similarly exposed human beings. To make matters more complicated, there are many examples of different species of laboratory animals exhibiting different responses to the same chemical exposure!

The 'nuts and bolts' of animal testing, and the problems of test interpretation and extrapolation of results to human beings, comprise one of the central areas of controversy in the field of chemical risk assessment. They shall be with us, in one form or another, for the remainder of this book. Suffice it to say at this point that animal tests are extensively used to identify the toxic properties of chemicals – in part because animals can be good models for humans and in part because we do not have other good choices – and will continue to be used for that purpose for a long time to come. We shall now begin to show how this is done.

Studying acute toxicity

We begin this exploration of the topic of animal testing by returning to the main subject of this chapter – acute toxicity. Systematic investigation of the toxic properties of a chemical usually begins with identification of what is technically called the acute lethal dose–50 (LD50): this is the (single) dose of a chemical that will, on average, kill 50% of a group of experimental animals.

The LD50 has become one standard measure of a chemical's acute toxicity. It is obtained by administering a range of doses of the chemical of interest to several different groups of experimental animals – a bioassay is used. The objective is to expose each group of animals to a dose sufficiently high to cause a fraction of them to die. A typical result, for example, might see 9 of 10 rats die in the group receiving the highest dose, perhaps 6 of 10 in the next highest group, 3 of 10 in the next, and 1 of 10 in the group receiving the lowest dose. Doses are typically administered using a stomach tube – the so-called *gavage* method – to allow the toxicologist an accurate quantitative measure of the delivered dose.

If this type of dose and response (in this case, the response is death) information is available, a simple statistical technique is applied to estimate the LD50 – the dose that will on average kill 5 of 10 animals, or 50% of the animals in any similar group were the test to be repeated.

Note that it would be possible to administer a dose sufficiently high to kill all the animals in every test group; that dose and every imaginable higher dose represents the LD100. But this type of information is not very useful to the toxicologist; the way in which lethality changes with dose and the point at which a single dose does not appear to have lethal potential are much more telling pieces of information.

Note also one other extremely important point; in the range of doses used, not every animal in each test group died. Only a certain fraction responded to the dose by dying, even though all animals in each test group are of the same species, sex, and even strain. Laboratory *strains* are members of the same species (rats, e.g.), that are very closely related because of genetic breeding; biologists thus refer to the Wistar strain of rat, or the Sprague-Dawley strain, or the Fisher 344 strain, and so on. These animals have been bred to achieve certain characteristics that are desirable for laboratory work.

The LD50, then, represents the dose at which animals have a 50% probability, or *risk*, of dying. This is our first specific example of risk information.

Identifying the LD50 is not the only purpose of the acute toxicity study. The LD50 provides a reasonably reliable indication of the relative acute toxicities of chemicals, and this is obviously important. But an even more important reason exists for conducting such tests, and that is to prepare the way for more extensive study of subchronic and chronic exposures.

During the acute toxicity determination the toxicologist carefully observes the animals for what are called 'toxic signs'.[7] If the animals appear to have difficulty in breathing, this sign indicates an effect of the chemical on the respiratory system. Tremors, convulsions, or hind limb weakness suggest the chemical damages the nervous system, or actually the neuromuscular system (nerves control muscle response). Redness or swelling of the skin points to a dermal toxicant. These types of observations help to identify the specific targets that the particular chemical may affect in tests of longer term duration, and the toxicologist can plan accordingly.

Knowledge of the range of doses that causes death also helps the toxicologist select doses for subchronic and chronic studies, which are the subject of the next chapter.

There are some other acute toxicity tests in which non-lethal outcomes are sought. These include studies of the amount of chemical needed to cause skin or eye irritation or more serious damage. Test

---

[7] Symptoms are what human patients can tell doctors about. An animal can't tell the toxicologist if it has a headache or an upset stomach. The toxicologist reads the 'signs'.

systems developed by J.H. Draize and his associates at the Food and Drug Administration in the early 1940s are still used for these purposes. Warning labels on consumer products are based on the outcomes of the Draize test.

The Draize test in particular has commanded much attention from animal rights activists, because it involves direct introduction of chemicals, typically consumer products of many types, into the eyes and onto the skin of rabbits. Ethical issues of several types arise in connection with the use of animals for toxicity testing. There are widely-accepted guidelines concerning the appropriate care to be given laboratory animals. These guidelines assume that animals can ethically be used for toxicity testing and other types of scientific and medical endeavors, but many in the animal rights movement question these premises, to greater or lesser degrees. These issues are important ones to toxicologists and a significant segment of the toxicology community is concerned to seek alternative means for acquiring some types of toxicity information. But these issues and their ramifications are outside the scope of this book.

6

Slow Poisons

'A toxicologist is someone who makes chemicals toxic'
(overheard by the author at a meeting of the
Society of Toxicology)

This remark is not far off the mark. In the last chapter we saw that the toxicologist is concerned to find the lethal dose of a chemical as the first step in its toxicological characterization. The dose is increased until the lethal range is reached. While in studies of subchronic and chronic duration the toxicologist does not want to push the dose high enough to cause the early deaths of the animals, he or she does seek to administer doses that will elicit detectable toxicity. The purpose of subchronic and chronic testing is to identify the types of adverse health effects produced by a chemical administered repeatedly, for large fractions of a lifetime, the dose at which toxicity begins to appear, and the manner in which toxicity changes above the minimum toxic dose. All substances, no matter how seemingly innocuous, can be made to produce some adverse effect if the dose is boosted to a high enough level. In this sense, all substances are toxic; they vary only in the nature of the toxic effects they produce and in the doses needed to produce those effects. Again, we observe that chemicals do not divide neatly into 'toxic' and 'non-toxic' categories.

The toxicologist also would like to identify the maximum dose at which the chemical produces *no observable toxicity*. The latter dose is termed the 'no-observed effect level' or NOEL, and it is used by the risk assessor to evaluate the likelihood of health damage in groups of people exposed to various doses of the chemical.

This chapter deals with 'slow poisons,' but this title, while conveying a message that has a popular meaning, is a little misleading. A more accurate title might be 'Identifying the Toxic Properties of Chemicals When They are Administered at Doses That Do Not Give Rise To

Immediately Observable Adverse Effects, but That Produce Effects Only After Repeated Dosing for Large Fractions of a Lifetime.' This statement avoids the false impression conveyed by a title that suggests there are two categories of chemicals, those that are acutely toxic and those that are 'slow poisons.' From the principles discussed so far, it should be obvious that all chemicals can be both 'fast' and 'slow' poisons, depending upon the size, duration, and other conditions of dosing. 'Slow poisoning' is perhaps a better title.

Slow poisoning can occur in several different ways. In some cases, chemicals or their metabolites may slowly accumulate in the body – rates of excretion are less than rates of absorption – until tissue and blood concentrations become sufficiently high to cause injury. Delayed toxicity can also be brought about by chemicals that do not accumulate in the body, but which act by causing some small amount of damage with each visit. Eventually, these small events, which usually involve some chemical interaction between the visiting chemical and normal cellular constituents, add up to some form of injury to the organism that can be observed and measured by the toxicologist.

Another possible mechanism of delayed toxicity involves creation of some serious form of cellular damage, involving the cell's genetic machinery, as a result of one or a very few chemical exposures. The damage may be passed on within the cell's genetic apparatus, to future generations of cells, even if the chemical causing the initial damage never again appears in the body. The reproducing, but deranged cells, if they survive, may eventually create a disease state, such as cancer, in the host. Whether this third type of delayed toxicity is associated with more than just a few chemicals is unknown, but, at least in theory, it is a possible mechanism of slow poisoning.

Knowledge of which mechanism of delayed toxicity is operating in specific cases can not usually be gained from the animal test; additional studies of ADME, and of interactions of the chemical with cellular components, are necessary to understand mechanisms of delayed toxicity. This chapter addresses mostly the ways slow poisoning is detected, the types of adverse effects that can appear, and some of the chemicals that can produce it. Mechanisms of toxic action come under review later, in Chapter 8.

The emphasis in this chapter is on the use of animal tests, or bioassays as they are sometimes called, to identify toxicity. Epidemiology has been important in uncovering the effects of subchronic and chronic exposures to some very important chemicals, including those capable of causing cancer. Carcinogens are a special breed of chronic toxicant and are given separate treatment in the next chapter. The methods of epidemiology will be reviewed in the chapter on carcinogens and will be given only scant attention in this chapter, where the emphasis is on the identification through animal tests of important, but non-carcinogenic, forms of toxicity.

Toxicity tests might be classified in two ways. The first type, referred to as general tests, are those in which groups of animals are dosed and various observations are made to determine how their health is affected. The second type is the specialized test in which the toxicologist is searching for certain forms of toxicity that are not easily observed with the general tests. In the latter category are tests to determine whether a chemical is deleterious to reproductive health, the developing embryo or fetus, the immune system, or even to behavior. Only very brief sketches of the specialized tests will be provided, because they do not teach much more about testing issues than can be learned from discussing the general tests. The next section will deal with toxicity testing, and then we shall discuss the important toxicity targets and some of the environmentally important chemicals that affect them.

Designing subchronic and chronic tests

Consider some new chemical that might be a useful pesticide. The manufacturer is required by federal law – the Federal Insecticide, Fungicide and Rodenticide Act – to develop all the toxicity data necessary for an evaluation of its risks to human health, prior to its commercial introduction. If the pesticide were to be approved ('registered', in EPA parlance) people are likely to be exposed to residues of the pesticide in certain foods. Workers might be exposed during manufacture and product formulation, or during application of the pesticide, and perhaps even when harvesting treated crops. Exposures might occur for only a few weeks each year for the workers, but could be fairly regular for the general population for a large part of a lifetime, if the product is commercially successful. The EPA is responsible for specifying the toxicity tests to be performed, the appropriate design of these tests, and the controls that the testing laboratory needs to exercise to ensure the integrity and quality of the test data. The EPA must receive and evaluate the test data (along with a great deal more data concerning pesticide usage, residue levels, environmental fate, and toxicity to non-human organisms, also developed by the manufacturer), and then find health risks to be negligible, before the pesticide can be registered and sold.

The manufacturer's toxicologists will begin testing with a determination of acute toxicity. If it appears the chemical is not unduly toxic, general subchronic and chronic, and a variety of specialized tests will be planned. A number of design issues need to be considered.

Route of administration

The general population will be exposed by ingesting pesticide residues on certain foods, so most of the tests will involve administration of the chemical mixed into the diets of the test animals. Some studies in which the chemical is put into a vapor or aerosol form will be conducted because some pesticide workers may inhale the substance. Skin toxicity studies are also needed because of possible pesticide worker exposures by this route. The principle here is straightforward: try to match the routes of exposure to be used in the tests to the likely routes of human exposure.

Test species

Most tests, both general and specialized, will be conducted using rats and mice. Dogs are used for certain specialized studies and rabbits for others. These species, specifically certain strains of them, have a long history of use as test subjects; their behavior in laboratory settings is understood and their dietary and other needs are well characterized. Another important consideration in the selection of test species and strains is knowledge of the types of diseases common to them. In the normal course of their lives certain diseases will 'spontaneously' develop in all species, so some of the animals assigned to the untreated, control groups in the toxicity studies will naturally develop certain disease conditions. (Many toxicologists use the unfortunate term 'spontaneous' when referring to diseases of unknown cause.) A classic example of this is the development of certain kidney diseases in elderly male rats. Knowledge of the normal range of rates of occurrence of various diseases in untreated animals is important to help toxicologists understand whether observed effects are chemically-induced or normal.

An issue of obvious importance in test species selection is the degree to which test results can be reliably applied to human beings. As we noted in the last chapter this is one of the principal problems in the evaluation of human risk, and we shall get back to it in the later chapter on risk assessment. For now, emphasis is on the selection of animal species and strains for their known reliability as experimental subjects. To put it in stark terms – the animals are used as toxicity measuring devices.

Controls

No study design is acceptable unless appropriate control animals are used. These are animals of the very same species, sex, age, and state of health as those to be dosed with the test chemical. The only difference between the controls and treated animals is the absence in the control group of exposure to the chemical to be studied.

Number of test subjects

If it is registered by EPA, millions of people might become exposed to the pesticide. Obviously it is not possible to use millions of laboratory animals in a test, and even thousands will present a logistical nightmare. For practical reasons, most tests are performed with 20 to 50 animals of each sex in each of the dose groups (a 'dose group' is a group of animals all of whom receive the same daily dose). Tests involving these numbers of subjects are obviously limited in some ways, and the toxicologist needs to consider these limits during the ultimate risk evaluation. But let us go into this matter now, because it is exceedingly important.

Suppose that, *unknown to the toxicologist*, a certain dose of a chemical causes a serious toxic effect – damage to certain brain cells – in one of every 50 exposed subjects. In other words, there is a 2 per cent risk of this form of toxicity occurring at our specified dose. Suppose this same dose is administered to a group of 50 rats, and the examining pathologist sees that one animal develops this particular form of brain damage. He also notes that none of the 50 untreated control animals develops the problem. Is it correct to conclude that the chemical caused this effect? The toxicologist finds a friendly statistician and is informed such a conclusion cannot be reached! Why not?

The statistician's role is to determine whether a disease rate of 1 in 50 is truly distinguishable from a disease rate of 0 in 50. The statistician will point out that there is only a very small chance (and chance, or probability, is what the statistician calculates) that this observed difference between the two groups of animals is actually due to the presence of the chemical in the diets of one of the groups, and its absence in the other. In fact, the statistician will state that not until the difference in disease rate is 0/50 vs. 5/50 is there reasonable probability that the observed difference is actually due to the chemical. In other words, a difference in disease rate of at least 10% (0/50 vs. 5/50) is necessary to achieve what is called a *statistically significant* effect. The difference necessary to achieve statistical significance will be smaller if more animals are used in the test (larger denominator), and larger if fewer animals are used.

Now it has been stated that, unknown to the toxicologist, the given dose of this chemical actually does cause a 2% increase in the rate of occurrence of this brain lesion. The problem is that, in an experiment involving 50 animals in a test group, the toxicologist cannot call an observed rate of 2% a true effect with any statistical legitimacy; indeed, not until the rate reaches an excess of about 10% can he conclude on statistical grounds that there is a difference in the responses of the two groups of animals. We are talking here about *what we can claim to know* with reasonable certainty as a result of experiments with limited numbers of subjects. If the numbers of animals in a test group are increased, lower disease rates can be detected, but rates very much below 5–10% cannot be achieved with groups of practical size.

What all this means – and this is of much concern in risk assessment – is that the animal tests we are describing cannot be used to detect excess diseases occurring with frequencies below 5–10%, and these are quite high risks, well above what we would deliberately tolerate in most circumstances (although pack-a-day smokers tolerate lifetime cancer risks about this high for themselves). The 2% excess risk in our example is also fairly high, but could not be detected in our experiment – it is a real risk, but remains hidden from us. The risk assessor has a means to deal with this type of limitation, and it shall be discussed in Chapter 9. For now, we simply restate that the numbers of animals assigned to toxicity test groups are largely determined by practical considerations, and that interpretation of test results needs always to consider the limitations imposed by the use of relatively small numbers of animals.[8]

---

[8] Cost is also an issue. Currently, a chronic feeding study involving two sexes of one species costs about $400 000. Two species are usually required. Inhalation studies are more expensive. Many studies in addition to chronic studies are required for new pesticide chemicals.

Before this topic is left behind, it should be noted that statistical significance is by no means the only consideration in interpretation of toxicity test results. If, in our particular case, the pathologist were to inform us that the particular brain lesion observed was extremely unusual or rare, we should certainly hesitate to dismiss our concerns because of lack of statistical significance. The toxicologist needs equally to understand 'biological significance,' and, in this case, would almost certainly pursue other lines of investigation (perhaps an ADME study to determine if the pesticide reaches the brain, or a toxicity test in other species) to determine whether the effect was truly caused by the chemical.

Dose selection

The usual object of test dose selection is to pick, at one extreme, a dose sufficiently high to produce serious adverse effects without causing the early deaths of the animals, and, at the other, one that should produce minimal or, ideally, no observable toxicity – a NOEL. At least one and ideally several doses between these extremes are also selected.

Some sophisticated guessing goes into dose selection. Knowledge of the minimum acutely toxic dose helps the toxicologist pick the highest dose to be used; it will be somewhere below the minimum lethal dose. There is usually little basis for deciding the lowest dose; it is often set at some small fraction of the high dose. Whether it is a NOEL will not be known until the experiment is completed. Sometimes bioassays have to be repeated to identify the NOEL.

A range of doses is desired to develop the all-important *dose–response* relationship: a quantitative portrait of the rate of toxic responses in the groups of test animals as a function of the size of the administered dose. In addition to rates of toxic responses, which are typically expressed as the fractions of exposed animals exhibiting the toxic effect, responses may also be expressed as the *severity* of the toxic injury. In some cases 100% of animals exhibit some form of toxic injury in response to a chemical exposure; what changes with changing dose is the severity of their injury. We shall see some examples of responses in which only a fraction of exposed animals exhibit an effect, with the fraction increasing with increasing dose; and others in which all animals respond at all doses, with the severity of the toxic effect increasing with increasing dose. Toxicologists refer to responses of the first type as *dichotomous* (a fancy term for 'yes/no'), and those of the second as *continuous*. The implications for risk evaluation of these two types of dose–response relations (there are also other types we shall not discuss), is one of the topics for Chapter 9.

Duration

The toxicologist usually moves from studies of a single exposure to ones in which animals are exposed on each of 90 consecutive days. The 90-day subchronic study has become a convention in the field. Rodents usually live 2–3 years in the laboratory, so 90 days is about 10% of a lifetime. An enormous amount of 90-day rodent toxicity data has been collected over the past several decades and has played a key role in judging the risks of environmental chemicals.

Our pesticide, we have said, will end up in certain foods and people could ingest residues over a portion of their average 70–75 year life span much greater than 10 per cent. So the toxicologist needs also to understand the toxicity associated with chronic exposures, about 2 years in rats and mice. The subchronic toxicity results are usually used to help plan the chronic study: to identify the doses to be used (usually a range of lower doses than those used subchronically) and any unusual forms of toxicity that need to be examined with special care.

One of the toxic effects the chronic study is designed to detect is cancer formation. Some toxicologists believe, in fact, that cancer is the only form of toxicity not detectable in 90-day studies! Indeed, it is difficult to find many examples of forms of toxicity occurring in chronic studies that were not detectable, at higher doses, in 90-day studies. It appears that, in most cases, the chronic exposure allows the effects that were detected in 90-day studies to be detected at lower doses, but does not reveal new forms of toxicity, except possibly cancer. This is not a sufficiently well-established generalization to support rejection of the need for chronic studies, and, of course, the toxicologist obviously needs to determine whether a chemical can increase the rate of tumor formation. So chronic studies will be around for some time.

Observations to be made

The extent, frequency, and intensity of observations to be made on the treated and control animals vary somewhat among types of study. In general, the toxicologist monitors at least the following parameters during the study:

- survival pattern
- body weights
- food consumption rates
- behavior patterns
- blood and urine chemistry

The animals receive a battery of clinical measurements, much like those people receive when they leave samples of blood and urine for testing after a medical examination. It turns out that body weight – reduced weight gain for growing animals or weight losses for adults – is a particularly sensitive indicator of toxicity. Its measurement does not provide much of a clue about the nature of the toxic effect that is occurring, but it is considered an adverse response in and of itself. In some cases it is due to reduced food consumption (and this is why food consumption is measured carefully), because the addition of the chemical to the diet makes the food unpalatable. In such a case, the chemical is obviously not producing a toxic effect; such a finding simply means that the experiment has to be repeated in a way that avoids the problem – in most cases this means undertaking the tricky task of introducing the required dose into each animal by stomach tube, the gavage method of dosing.

At the end of the study animals remaining alive will be killed, and examined by a pathologist. So will any animals that die during the course of the study, assuming their deaths are discovered before their tissues have begun to decompose.

The pathologist will first visually examine each animal inside and out. About 30 different tissues and organs will be taken and prepared for examination under a microscope – the so-called histopathological examination. As in the case of the pathologist who looks at tissues from people, the animal pathologist is characterizing the disease state or type of injury, if any, to be found in particular tissues. The pathologist often does this 'blind' to the source of the tissue, that is, without knowledge of whether the tissue came from the treated animals or the control animals.

So the animals used in the toxicity test, both treated and control animals, are subjected to extremely thorough 'medical monitoring,' even to the point of sacrificing their lives so that the toxicologist can learn in minutest detail whether any of their tissues have been damaged. Obviously, there's no way such thorough information could ever be collected from any imaginable study of humans exposed to chemical substances.

Conducting the bioassay

Once the design is established, a protocol will be prepared. All the critical design features and the types of observations to be made, and even the statistical methods to be used to analyze results, are specified in advance. Toxicologists, chemists, pathologists and statisticians are typically involved in drafting the protocol.

Some aspects of the mechanics of testing deserve mention. First, animals to be put into the control group and to the groups to be treated with the chemical need to be assigned in a completely random fashion. The animals are usually selected for testing not long after weaning, while they are still in a growing phase, and care must be taken to avoid any discrimination among the groups with respect to factors such as weight – the person assigning the animals to various groups should have a procedure to allow completely random selection of animals.

A second factor concerns the purity of the diet and water received by the animals. Careful chemical analysis is needed to ensure the absence of significant amounts of highly toxic chemicals, such as aflatoxin, metals such as lead, arsenic, or cadmium, or certain pesticides, that may be present in water and various feed ingredients.

If the oral route is to be used, the chemical may be mixed with the diet, dissolved in drinking water, or delivered by a tube to the stomach (gavage). An inhalation exposure requires special equipment to create the desired concentrations of the chemical in the air to be breathed by the animal. In any case, the analytical chemist must be called on to measure the amounts of the chemical in these various media after they have been added to guarantee that the dose is known with accuracy. Some chemicals decompose relatively quickly, or errors are made in weighing or mixing the chemical to achieve the desired diet, water or air concentrations, so chemical analysis of these media is essential throughout the study.

There are many other features of toxicity studies that require careful monitoring and record-keeping, but they won't be mentioned here. Suffice it to point out that conducting a chronic toxicity study requires extremely careful control and monitoring. Indeed, a series of discoveries during the 1970s of poor record-keeping, sloppy animal handling and, in a few cases, deliberate recording of false information in study reports led to the promulgation of federal regulations concerning 'Good Laboratory Practices'. The regulations specify the type of data collection and record-keeping and additional study controls that must be documented for studies whose results will be submitted to federal regulatory agencies. It is foolhardy these days to conduct toxicity tests in laboratories that can not demonstrate strict adherence to GLPs (note we are referring to toxicity tests, as distinct from toxicity research).

Study evaluation

At the end of the toxicity test the toxicologists and pathologists list all the observations made for each individual animal in each dose group. Analysis of these results is needed to identify effects caused by the chemical under study. All observations – body weights, clinical measurements, histopathology, and so on – have to be included in the evaluation.

Two types of analysis are needed: one concerns the biological significance of the results and the other their statistical significance.

The statistician is interested in determining the chance, or probability, that the rates of occurrence of certain injuries in specific tissues are different from the rates of their occurrence in the same tissues of control animals. Observations of numerical differences are not sufficient to conclude that the difference is due to the chemical treatment and not to simple chance. The statistician has several well-established techniques to estimate the probability that an observed numerical difference in the occurrence of a particular effect between a group of control animals and a group of treated animals is due simply to chance. It is conventional in many areas of biological science, including toxicology, to conclude that an observed difference is a 'real' one if there is less than a 1 in 20 probability that chance is involved. This is a scientific convention, not a law of nature. More or less strict criteria for statistical significance can be applied: the less the probability that we have observed a chance occurrence, the greater the probability that the chemical was responsible for that occurrence.

So the statistician does this type of analysis for each treatment group versus the control. When this is done the toxicologist can determine whether any of the observed effects are 'statistically significant' – whether and to what degree of confidence can the observed effect be said to have been caused by the chemical treatment.

The toxicologist looks not only at the rates of occurrence of various adverse effects, but also examines the question of whether the severity of certain forms of injury is significantly greater in treated animals. 'Severity' is not always quantifiable – it is in the eye of the pathologist.

So some judgment beyond what the statistician can offer through an objective analysis is always necessary to complete an evaluation.

Statistics do not tell the whole story. Not infrequently effects are seen at very low frequencies (not at statistically significant rates) that the toxicologist may think important and likely due to the chemical. The typical case involves the appearance of very rare or highly unusual diseases or forms of injury in treated animals – diseases or injuries that historically have never been observed, or observed only at extremely low frequencies, in untreated animals.

Test interpretation is often made difficult because diseases occur with very high frequency in specific organs of untreated animals. If an average of 80% of aged, untreated, male rats of a certain strain normally suffer from kidney disease, then it becomes difficult to determine, both statistically and biologically, whether any observed increase in this same disease in treated animals is truly related to the chemical. The high background rate of the disease obscures the effect of the test chemical. The same type of problem plagues the epidemiologist. Searching for specific causes of diseases such as breast, lung or colon cancers, that occur with relatively high frequency in human populations, is extremely difficult.

In the end the specific toxic effects that can be attributed to the chemical with reasonable confidence can be isolated from those that can not be so attributed. As in most areas of science, there will almost always be effects falling into an ambiguous zone, or there will be effects that are of uncertain significance to the health of the animal. In the latter category are, for example, minor changes in the rates of occurrence of certain normal biochemical processes, typically at the lowest test dose, unaccompanied by any other sign of disease or injury. The toxicologist simply does not know whether such a change – which is clearly caused by the chemical treatment – has any adverse consequences for the health of the test animal. In such cases there is uncertainty whether the lowest dose is a true NOEL; toxicologists can argue endlessly over such observations, but usually can't think of any good way to resolve the argument.

Specialized tests

So far the toxicologist has not made inquiries regarding a number of potentially important questions. Can the chemical hurt a developing embryo and fetus, perhaps producing birth defects? Can the chemical reduce male or female fertility, or otherwise impair reproduction? Can the chemical injure the immune system, or alter behavior? Can it cause cancer or mutations?

These sorts of questions can not be thoroughly explored with the general tests discussed so far. These tests do not, for example, provide for any mating of the animals. They do not allow chemical exposure of the females when they are pregnant. The chronic study may pick up an excess of tumors, but sometimes special chronic tests – called *cancer bioassays* – are performed to test the carcinogenicity hypothesis (Chapter 7). Studying the effects of chemicals on behavior, on certain neurological parameters, on the immune system, and on the materials of cellular inheritance, requires test designs and measurement techniques that are quite different from those needed for generalized testing, and which can not readily be built into the general tests. Moreover, many specialized tests are still in the developmental stage, and regulatory agencies tend to be reluctant to require their use until they have undergone validation of some type.

ADME studies are typically included in the battery of tests used to characterize the toxicity of chemicals, as well as other studies designed to trace the underlying molecular and cellular events that lead to toxicity. These studies of toxic mechanisms take many forms, and are better viewed as research studies; no general characterization of them will be made here, but some of the types of things such studies can reveal to aid understanding of risk will be mentioned at appropriate places in the remaining sections of the book.

Some slow poisons and their targets

Some specific examples of chemicals causing toxicity at various targets will be introduced at this point. Most can produce 'slow poisoning,' but the discussion also includes some acute responses as well. The toxic properties of the chemicals to be discussed have been learned by the types of general and specialized animal tests just discussed. In many cases they have also been learned from epidemiological studies and case reports. Carcinogens, as we have already mentioned, are excluded until the next chapter.

There are several possible ways to categorize chemical toxicity. Perhaps the most common is by grouping chemicals according to the targets they can damage, and it is the approach followed here.

One result of this approach is that, from the chemist's point-of-view, the grouping is a hodge-podge. Thus, under liver toxicants, are grouped a number of organic solvents, some metals, certain pesticides, some naturally-occurring chemicals, a few pharmaceutical agents, and a miscellaneous collection of industrial chemicals of diverse structural properties. In fact, each target group will contain such an assortment. What is seen when toxicity is grouped by target is that substances of diverse structural, chemical, and physical properties can affect the same biological target. And although chemicals having highly similar structures tend to produce similar forms of toxicity, there are many dramatic examples in which small modifications in chemical structure can result in substantial shifts in toxic potency and even toxicity targets.

In a few cases, the way a chemist might group chemicals does match, to a degree, the way the toxicologist groups them. Chemicals that can dissolve fatty materials, and that are used as solvents for them, tend to have similar physical properties. Many of these chemicals can impair the nervous system by a biological mechanism that depends upon their characteristics as solvents. But this type of matching is not the general rule.

The survey that comprises the remainder of this chapter is highly selective. The target groupings included by no means encompass all of the targets that can be affected by chemicals. Omitted is much mention of the cardiovascular system, the eye, and the various glands that secrete hormones; the immune system has also been given short shrift. These omissions should not be taken as suggesting that chemical toxicity is not important for these targets.

The three targets that are the first point of contact between environmental chemicals and the body will be discussed first: *the gastrointestinal tract, the respiratory system,* and *the skin.* Recall from Chapter 3 that chemicals enter the *blood* after absorption, so this fluid is the next target (Figure 1). Then comes the *liver,* the *kidneys* and the *nervous system.* The chapter concludes with a discussion of some chemicals that can damage the *reproductive system* and some that can cause birth defects, the so-called *teratogens.*

Only a few, well-known examples of chemicals that can damage these targets are presented; for most of these targets a complete list would include several dozen up to several hundred substances! Notice also that some chemicals appear on two or three lists. The chemicals reviewed were selected primarily because they provide good illustrations of various toxicological phenomena, and not necessarily because they are environmentally important (though some certainly are).

Keep in mind that because a chemical is listed as producing toxicity does not mean it produces this toxicity under all conditions of exposure. Whether a liver toxicant is likely to produce its effects in human beings under their actual conditions of exposure is only partially answered by knowledge that it has been shown to cause liver toxicity in test systems or in certain groups of highly exposed people. A full risk assessment is needed to answer the ultimate question, and a great deal more must be known before that question can be dealt with.

A final note of caution: the toxicity information provided on individual chemicals is by no means complete. Data have been selected to illustrate certain principles; no attempt is made to provide anything close to a thorough toxicological evaluation of any of the chemicals discussed. The list of Recommended Readings, appearing after the final chapter, lists several sources of toxicity information on individual chemicals.

Respiratory system

The various passages by which air enters the body, together with the lungs, comprise the respiratory system. Its principle purpose is to move oxygen into the blood, and to allow the metabolic waste product, carbon dioxide, to exit it and to leave the body in exhaled air. The exchange of these two gases occurs in the lung. The respiratory system serves other purposes as well, and includes a mechanism for the excretion of toxic chemicals and their metabolic products.

The so-called respiratory tract has three main regions. The naso-pharyngeal region includes the nasal passages and the pharynx, which is a cavity at the back of the mouth; these passages are the first point of contact for air and chemicals carried in it, in the form of particles, gases, or vapors. Below the pharynx lies the second, tracheobronchial region, which includes the trachea (windpipe), at the top of which sits the larynx (voice box); off the windpipe extend two tubes called bronchi, one leading to each lung. The bronchi undergo several branchings within the lung and finally lead to the *pulmonary* region. Here the branches lead to bunches of tiny air sacs, which in turn end in small 'pouches' called alveoli, where oxygen entering the body and exiting carbon dioxide change places. There are about 500 million alveoli in the adult human being, with a total surface area of about 500 square feet! The two bronchi and their many branches can be thought of as inverted trees extending into the lungs, and the air sacs ending as alveoli as leaves on those trees. Although the air within these respiratory structures is ostensibly 'inside' the body, in fact it is not; all of it is connected without obstruction to the air outside the body, and components of air, including oxygen, get truly inside the body – into the bloodstream – only after they pass through the cells lining the alveoli.

Gases, vapors, and dust particles can move into the airways from the environment and penetrate to the pulmonary region in several ways. The gases and vapors can readily move through the three regions, but dust particles are blocked at several steps along the path. Large particles, those typically sneezed out, do not get beyond the nasopharynx. The trachea and bronchi are lined with epithelial cells (*epithelial* is the adjective biologists attach to cells that act as linings within the body), that secrete mucus and that also hold little, hairlike attachments called cilia. The cilia and mucus can collect particles that are small enough to negotiate their way through the nasopharyngeal region, and the cilia move those particles up to the mouth, where they collect in saliva, either to be excreted or swallowed. Some very tiny particles, generally those less than one micrometer (μm) (one-millionth of a meter) in diameter, instead of being caught in the 'mucociliary escalator', as it is called, manage to get eaten up or engulfed by cells called phagocytes. Phagocytes carry dust particles into various lymph nodes, whereby they can enter the blood. Gases and vapors can, to varying degrees that depend upon their chemical and physical properties, be absorbed into the blood at any of the three regions of the respiratory tract, but most absorption takes place in the pulmonary region. Some of these substances cause systemic toxicity, others only cause local toxicity in the respiratory tract; others cause both types of toxicity.

Toxicologists who study the responses of the respiratory system to foreign chemicals generally categorize those responses according to their biological and pathological characteristics.

Irritation is caused by many chemicals, including the common gases ammonia, chlorine, formaldehyde, sulfur dioxide, and dust containing certain metals such as chromium. The typical responses to a sufficiently high level of such substances is constriction, or tightening, of the bronchi; this is accompanied by *dyspnea* – the feeling of being incapable of catching the breath. With the airways constricted in this fashion, oxygen can not get to the pulmonary region at a fast enough pace to satisfy the body's demands. This type of constriction brings on asthma attacks and, if chronic, then a long-lasting bronchitis, or inflammation of the bronchi, may ensue. Sometimes a serious swelling, or *edema*, occurs in the airways, and when irritation is especially serious, it can pave the way for microorganisms to invade the tissue of the airways and lungs, to cause an infection. A lethal exposure is one that completely overwhelms the responsive power of the respiratory system, and turns off the respiratory process for good. Generally, however, irritation and edema subside following cessation of exposures – they are often reversible phenomenon.

Exposures to irritating levels of gases such as ammonia, chlorine, and hydrogen chloride, and metals such as chromium, usually occur only in certain occupational settings, although occasionally the general population becomes exposed because of a transportation or industrial mishap (although solutions of ammonia are sold as household products, and almost everyone has experienced in a mild way the pungent qualities of ammonia gas). Sulfur dioxide is a somewhat different matter. This gas, SO_2, is produced by burning of fossil fuels, all of which contain some sulfur. Metal smelting operations also produce SO_2 as a by-product, because of the presence of sulfur in the raw ores. These burning and smelting processes also produce particles containing SO_2 reaction products, called sulfates. As a result, sulfur dioxide and fine particles containing various sulfates are common air pollutants to which millions are exposed on a daily basis. At some times and in some places levels can shoot up and cause disturbingly unpleasant irritating effects, as well as bronchoconstriction. An especially serious pollution event is the so-called 'sulfurous smog' caused by the accumulation of dense particles containing sulfuric acid, H_2SO_4. The most serious of such smogs occurred in London in 1952, where about 4000 people died because of the event. Twenty deaths occurred in 1948 in Donora, Pennsylvania, because of sulfurous fog.

People who suffer from other pulmonary diseases that interrupt the flow of oxygen are especially sensitive to the irritating effects of SO_2 and its particulate derivatives. This gas and several other gaseous air pollutants, to be mentioned in a moment, can cause other, delayed toxic effects in the respiratory system. Note also that these same chemicals are the principal causes of acid rain.

Some especially irritating organic chemicals are certain gaseous or highly volatile compounds called aldehydes, the most well-known of which is formaldehyde. This gas is a natural product of combustion, and is present in smoke, including that from tobacco. It and a few related aldehydes are the principal agents causing irritation in the upper respiratory region when smoke is inhaled. Formaldehyde is, of course, a major industrial chemical (6 000 000 tons produced in 1987) that is used to manufacture plastics, including an insulating material, urea-formaldehyde foam, that has been installed in millions of homes; residual levels of the gas may emanate from those materials and be irritating to some individuals. The chemical is also used to manufacture 'sizing' for synthetic fabrics, a process that gives a permanent press to certain clothing. Formaldehyde odors are common in stores specializing in clothes of this variety. Because of its natural occurrence and industrial production formaldehyde is omnipresent in indoor and outdoor atmosphere.

$$\overset{\displaystyle O}{\underset{\displaystyle H-C-H}{\|}}$$

Formaldehyde

A second category of respiratory toxicity is that characterized by damage to the cells anywhere along the respiratory tract. Such damage can cause the release of fluid to the open spaces of the tract, and result in accumulation of that fluid, or edema, in several areas. These edematous reactions can occur after an acute exposure to some chemicals, although the production of edema can be delayed, and arise after subchronic and chronic exposures.

Two common and widespread air pollutants, ozone (O_3), a potent oxidizing agent, and nitrogen dioxide (NO_2), are good examples of chemicals that can cause cellular damage in the airways and lungs. Sustained exposure to these gases can cause *emphysema*, with accompanying loss of capacity for respiratory gas exchange; this condition can, of course, lead to serious physical disabilities. The effects of these gases are compounded by smoking.

The Clean Air Act recognizes a number of so-called 'primary air pollutants,' and the EPA has established standards for these substances. Ozone, nitrogen dioxide, and sulfur dioxide are among these (the others are carbon monoxide and lead, discussed below, and 'total suspended particulates'). EPA's standard for ozone is 0.12 parts of the gas per million parts of air (0.12 ppm), as a one-hour exposure limit that is not to be exceeded more than once yearly. Nitrogen dioxide's limit is 0.05 ppm as an annual average. These standards are designed to prevent chronic respiratory toxicity of any type.

Some occupational situations, if inadequately controlled, can create opportunities for similar damage to the respiratory system. Exposures to certain forms of the metals nickel and cadmium, ordinarily as airborne particulates, can cause cellular damage, edema, and, if sustained for sufficiently long periods, emphysema. Many other metals, usually only in certain of their many chemical forms, can produce emphysema upon subchronic or chronic exposures.

A particularly interesting example of a respiratory toxicant is the pesticide called paraquat. This organic chemical produces serious and generally delayed pulmonary edema after *ingestion*. This example illustrates the phenomenon of systemic toxicity – toxic effects at sites of the body distant from the site of initial contact and which can be reached only after a chemical enters the bloodstream – and further that the lungs can be not only a site of direct, local toxicity, but can also be a target for systemic effects. Paraquat after being absorbed through the gastrointestinal tract, enters the circulatory system and thereby reaches an organ where it is particularly active, the lungs.

Fibrosis is a third category of pulmonary damage. Certain particles and dusts, when inhaled for long periods in sufficiently tiny sizes, can create cellular damage in the lungs of a type that causes those cells to exude fibrous materials, much like tiny filaments of connective tissue or gristle. If there's a sufficient build-up of these fibers, the lung tissue can become rigid and lose function. In its advanced form fibrosis is a serious and debilitating disease.

Fibrosis was first recognized in certain occupational settings. One of the well-known conditions of this type is silicosis, which is brought about by long-term, uncontrolled exposure to certain crystalline forms of silica (SiO_2) and certain related substances called silicates. These minerals are widespread on earth, in fact most of the inorganic, non-aqueous earth consists of silica and silicates. Many of these minerals (e.g., quartz) have major industrial uses. It is important to emphasize that silica and silicates occur in both crystalline and non-crystalline (amorphous) forms, and it is only the former that causes silicosis. Occupational exposures to crystalline silica and silicates have to be carefully controlled.

Asbestos is the name given to several different fibrous forms of silicates; these go under names such as crocidolite, amosite, and chrysotile. These forms are, to varying degrees, also capable of eliciting fibrosis in lung tissues. In Chapter 7 it will be seen that some forms are capable of more serious damage – cancers of the lung and the mesothelium.

The final category of respiratory response worth noting includes some allergies. Allergic reactions are a special brand of adverse effect, resulting from the body's response to many types of foreign agents, including microorganisms, certain large protein molecules, and even some relatively small foreign molecules. The first exposure to an antigenic chemical may result in the interaction of that chemical with certain normal proteins to form complexes called antigens; antigens in turn provoke the formation of other complex entities called antibodies, which remain in the body. No allergic reaction, but only those biochemical changes, takes place as a consequence of this first exposure.

But subsequent exposure to the same chemical can for some people be quite devastating. This exposure results in an interaction between newly formed antigens and the antibodies that were produced from the first exposure, which in turn elicits a series of biochemical and physiologic responses ranging from mild flushing of the skin all the way to death. One chemical that produces allergic-like responses in the lungs is toluene diisocyanate (TDI), a volatile chemical used to produce polyurethane plastics. It is a major industrial chemical, and worker exposure needs to be tightly controlled. The mildest manifestation of a TDI-induced allergenicity is bronchoconstriction, but at sufficiently high exposures some sensitive individuals can suffer substantial losses in pulmonary function from which they only slowly recover.

The allergic response differs from the usual toxic response in that a prior exposure is necessary to create the conditions for an adverse response.

Gastrointestinal tract and skin

The two other body areas that are first contact points for environmental chemicals are the gastrointestinal tract and the skin. The GI tract is that long and many-faceted tube beginning at the mouth and extending downward as the pharynx and the esophagus, then enlarging as the stomach, narrowing again as the small intestine, and ending as the large intestine (which consists of the cecum, colon, rectum, rectal canal and anus). The GI tract is, as shown in Chapter 3, a major point of entry for environmental chemicals present in food and water, and even in soils and dusts. It is relatively rare that environmental chemicals reach sufficiently high levels in those media to produce significant toxicity directly in the GI tract, although accidental ingestion of many chemicals may cause severe injuries to it. Highly caustic materials such as lye (sodium hydroxide) have been accidentally ingested by many individuals and have been shown capable of causing serious damage to the lining of the GI tract. Such materials, because of their strongly alkaline properties, essentially destroy the natural fatty chemicals present in cell membranes, either of the GI tract, or at any other site of direct contact. But doses below those causing such readily noticeable effects appear not to cause any adverse reactions upon subchronic or chronic exposure.

While most cases of GI tract distress are probably due to microbiological agents or their toxins, many chemicals are capable of inducing vomiting and diarrhea and other GI tract responses. Poisoning with heavy metals such as lead, cadmium, and arsenic may be suspected in patients reporting with severe abdominal pain, nausea and vomiting, although these are typically the consequence of acute, high level exposures. In many cases the GI tract response leading to vomiting or diarrhea is an indirect effect of the chemical, secondary to a systemic attack on the nervous system which controls GI tract behavior. In contrast, certain microbiological agents, such as species of Salmonella present in contaminated food or water, may grow in the GI tract and directly induce these effects.

Perhaps the most important toxicological role played by the GI tract is its influence over the absorption of chemicals entering it. That absorption rates vary widely among chemicals has been explained in Chapter 3, but how the GI tract and its contents contribute to this phenomenon was not noted. Mechanisms of absorption are many and varied, and are influenced by the type and quantity of food present at the time of chemical ingestion, the pH (degree of acidity) of various portions of the GI tract, and even the nature and activity of the microorganisms that normally live in the intestines. In fact, metabolism of certain chemicals brought about by these microorganisms can play a crucial role not only in their absorption but also in the nature of the systemic toxicity they ultimately produce.

Skin is our largest organ. The average adult's body surface area is about nine square meters and the skin that comprises it weighs 20–30 pounds (about 15% of body weight)! All sorts of chemical agents come into direct contact with it. A few are metabolized in parts of it, and many pass through it into the circulation.

Because adverse skin responses are so easily recognizable, this organ was among the earliest subject to scrutiny, mostly by physicians interested in occupational diseases. Bernardino Ramazzini's tract of 1700, *De Moribis Artificum Diatriba*, contained many examples of skin diseases associated with occupational exposures, and, as noted earlier, the seminal work of Percival Pott on occupationally-induced cancers, published in 1775, revealed the role of soot in the production of cancers on the skin of the scrotum in London chimney sweeps.

Irritation of the skin is brought on by a very large number of chemicals. It is characterized by reddening, swelling, and itching, which generally subside after exposure ceases. Allergic responses, as in the case of those occurring in the respiratory system, require prior sensitization; subsequent exposures bring on an attack. Formaldehyde causes allergic responses in the skin of sensitive individuals at exposures lower than those necessary to elicit irritation.

No doubt the greatest environmental threat to the skin is not chemical, but is rather a physical agent, sunlight; most skin cancers are caused by excessive exposure to ultraviolet radiation from the sun.

Blood and lymphatics – hematoxicity and immunotoxicity

Nutrients absorbed from the gastrointestinal tract and oxygen from the lungs enter the blood stream and are thereby carried throughout the body, where they feed the machinery of cells. Blood is, of course, pumped by the heart through the *circulatory system*, a complex network of tubes, called vessels. There are three main types of blood vessels: the *arteries* carry blood containing nutrients and oxygen away from the heart and the *veins* carry blood and cellular waste products including carbon dioxide back to it; there is also present a system of small, thin-walled tubes called *capillaries* which branch off the arteries and subsequently merge to form veins.

Blood *plasma* is the liquid, mostly water, that carries several types of blood cells as well as nutrients, and other chemicals, such as hormones, that need to be transported around the body.

In Chapter 3 we explained how chemicals foreign to the body can enter the circulatory system and be transported to various parts of the body. If the concentrations of these substances or their metabolic products reach sufficiently high levels, systemic toxicity can result. Different chemicals affect different organs and systems of the body because of differences in the rate and manner of their absorption, distribution, metabolism, and excretion. Some chemicals are directly toxic to the elements of the blood. Others bring about changes in certain elements of the blood that become detrimental to other systems of the body.

The most well-known example of the latter is carbon monoxide. This simple gas (CO) is a product of incomplete combustion of organic substances. Complete combustion results in the conversion of carbon-containing chemicals, such as the hydrocarbons used as fuels, to carbon dioxide (CO_2), but because most combustion systems can not allow for the presence of all of the oxygen needed, some of the carbon ends up in the less oxidized form of CO. Incomplete combustion of gasoline in trucks and automobiles is one of many sources of the gas. Individuals who cannot escape from fires can be exposed to very high levels of CO and many fire deaths are related to this gas. Firefighters, garage workers and traffic policemen can experience relatively high concentrations while on the job, and all the rest of us inhale the gas throughout the day because of its ubiquitous presence in the atmosphere. Smokers get an additional dose. CO gradually oxidizes to CO_2 and so does not continue to accumulate in the atmosphere.

The gas has a molecular size and shape similar to oxygen (O_2). When oxygen passes from the lungs into the blood it interacts with a large molecule called *hemoglobin* (Hb). This vital chemical is present in the red blood cells (*erythrocytes*). In addition to a large protein component Hb contains a complex organic compound called heme; the heme molecule carries within it an ion of the inorganic element iron. Under normal circumstances oxygen molecules, after they pass through the lungs, interact with Hb, specifically with the iron-heme portion.

$$Hb + O_2 \rightarrow O_2Hb$$

The O_2Hb molecule is called oxyhemoglobin and has a bright red color. The red blood cells transport O_2Hb to all the cells of the body, where O_2Hb dissociates, yielding up the needed O_2 molecules.

Carbon monoxide is dangerous because, like oxygen, it has an affinity for Hb, and produces a bright, cherry-red substance called carboxyhemoglobin.

$$Hb + CO \rightarrow COHb$$
Carboxyhemoglobin formation

In fact, CO's affinity for Hb is even greater than that of oxygen, by several hundred times! Because the body's supply of red blood cells and Hb is limited, the presence of CO in inhaled air can deprive the body of oxygen, a condition called anoxia. The nature, duration, and severity of resulting toxicity depend upon the blood COHb level created which, of course, depends upon the concentration of the CO in the inhaled air and the length of time the air is inhaled. The presence of COHb in the capillary blood imparts an abnormal red color to skin and fingernails. The conditions creating toxicity arise in the blood; the actual effects appear in the nervous system, in the heart, and elsewhere.

One of the most carefully worked-out dose—response relationships is that for carbon monoxide poisoning. Based on controlled studies of exposures in humans at low levels and on observations in humans who have suffered high level exposures because of occupation or because of accidents or suicide attempts, the relationship between blood levels of carboxyhemoglobin (COHb) and toxicity is understood as follows:

Per cent COHb in blood	Signs and symptoms
0–10	*See text, below
10–20	Headache
20–30	Headache, throbbing in temples
30–40	Headache, dizziness, nausea, vomiting, dimness of vision
40–50	Collapse, increased pulse rate
50–60	Increased respiration, coma
60–70	Coma, convulsions, depressed heart rate
70–80	Respiration severely depressed, death within hours
80–90	Death within one hour
90–100	Almost immediate death

Except for death and possibly damage to the heart, these effects are reversible when the CO source is removed, because COHb eventually dissociates and releases CO to the lungs, where it can be excreted. The treatment for CO poisoning is administration of oxygen to hasten dissociation of the COHb molecule.

No human health effects have been detected at COHb blood levels below about 2 per cent ('background' levels in non-smokers average about 0.5%). Subtle effects on the nervous system, such as reduced ability to sense certain time intervals, have been reported at blood levels of 2.5%. At COHb levels of 5% certain cardiovascular changes are detectable, especially in patients with coronary heart disease. Heavy smokers exhibit COHb levels in the range of 5–6 per cent, and if they happen to be pregnant, the fetus can suffer the effects of oxygen deprivation.

The quantitative relations between blood levels of COHb and air levels of CO have been well worked out. In general, the blood level achieved is a function of both air concentration and the length of time the individual breathes the air ($C \times T$). Legal limits on workplace and environmental concentrations have been established to avoid significant COHb levels in the blood, but of course this goal is not always realized.

The hemoglobin molecule can be adversely altered in other ways by certain chemicals. The ion of iron that is present can be oxidized by certain chemicals to produce a brown-black compound called *methemoglobin*. The latter cannot bind to oxygen, so chemicals creating it can produce serious toxicity if the concentrations generated are sufficiently high. Nitrites, inorganic ions having the structure NO_2^{-1}, are particularly successful at creating methemoglobinemia (excess methemoglobin in the blood).

In addition to erythrocytes, blood contains white blood cells, called leukocytes, of several types, and platelets, also called *thrombocytes*, which control blood clotting. Hematopoiesis (from the Greek, 'haimo', for blood, and 'poiein' for 'to make') is the process by which the elements of the blood are formed. The marrow of bone contains so-called stem cells which are immature predecessors of these three types of blood cells. Chemicals that are toxic to bone marrow can lead to *anemia* (decreased levels of erythrocytes), *leukopenia* (decreased numbers of leukocytes), or thrombocytopenia. *Pancytopenia*, a severe form of poisoning, refers to the reduction in circulatory levels of all three elements of the blood. One or more of these conditions can result from sufficiently intense exposures to chemicals such as benzene, arsenic, the explosive trinitrotoluene (TNT), gold, certain drugs, and ionizing radiation. Health consequences can range from the relatively mild and reversible to the severe and deadly. Some chemicals produce an excess of certain of the blood's elements and this may signal equally serious consequences for the health of the affected person. Such is the case with the various *leukemias*, characterized by greatly increased numbers of certain leukocytes. Benzene, a chemical of substantial environmental importance, and certain drugs have been associated with the production of certain types of leukemias in humans (Chapter 7).

The tiny blood vessels called capillaries run close to the individual cells that comprise the various organs of the body, but do not touch them. Instead the nutrients, oxygen, and foreign chemicals they carry migrate into certain tissue fluids (called intercellular, or interstitial, fluid) which surround and bathe cells. This fluid provides the contact between the circulatory system and the body's drainage system, called the *lymphatics*. Intercellular fluid carries nutrients, oxygen, and chemicals to cells, and carbon dioxide, organic waste (including the chemicals or their metabolites) away from them. Some of these wastes enter the capillaries that combine to reenter the veins, and the rest (particularly waste molecules too large to enter venous capillaries) pass into lymph.

The lymphatic system consists of vessels and various small organs. The lymph nodes (or glands) are found at several locations along the system of lymph vessels, often bundled into groups (as in the armpits). The glands produce one class of white blood cells called *lymphocytes*, cells that produce the body's 'defense proteins,' called antibodies. Other lymph glands include the spleen, the tonsils, and also the thymus. The last mentioned is located in the upper region of the chest and wastes away − atrophies − after puberty. Lymph glands are all capable of trapping foreign bodies such as proteins and bacteria; they are main lines of defense against infections ('swollen glands', which can be detected by feeling areas of the body where the nodes group together, are indications that the body has been invaded by infectious agents).

The liquid carried in lymph vessels is called lymph; it contains lymphocytes and substances acquired from intercellular fluids. Lymph is a key disposal system for the body's wastes, including the debris from infectious agents and foreign proteins trapped in it.

The lymphatics are intimately involved in the several complex processes whereby the body protects itself from foreign agents − the immune system. A rapidly evolving discipline is that called immunotoxicology, the study of the adverse effects of chemicals on the components and operations of the immune system. The consequences of exposure to immunotoxic agents range from suppression of immunity, which can lead to reduced resistance to infection and certain diseases including cancer, to mild allergic responses. Some important immunosuppressive agents include benzene, PCBs, and a variety of therapeutic drugs. Chemicals capable of producing allergic responses include toluene diisocyanate (TDI), a very important commercial product used to make certain plastics and resins that we mentioned above in connection with allergic responses in the respiratory system, and certain metals such as nickel. Animal models to study immunotoxicity are still under development and are not routinely used for testing new chemicals; some aspects of immune system toxicity can be detected using conventional test designs, but specialized tests are needed for a thorough evaluation. Although some forms of alteration of the immune system are clearly detrimental to health, there is still debate about the relevance to health of certain immune system changes that can be induced by chemical exposures. It is not clear that all biological changes brought about by a chemical will threaten health. There are vast areas of immunotoxicity waiting to be explored, and it is certain to become an increasingly important component of the toxicological evaluation of environmental chemicals.

Liver – hepatotoxicity

Liver damage produced by toxic chemicals is called *hepatoxicity*, and has been under study for a very long time; this large and interesting organ has perhaps been subject to more extensive examination by toxicologists than any other. Among many others the liver plays a key role in digestion, regulation of blood sugar levels, storage of vitamins and iron, and synthesis of proteins and other essential molecules. The cells of the liver, called *hepatocytes*, are marvelously intricate and efficient chemical factories, and they contribute greatly to normal metabolism and to the metabolism of foreign chemicals.

This organ may appropriately be called a gland because it secretes a complex substance called *bile*. (The term 'gland' is applied to any organ that secretes substances, typically but not only hormones; another gland having a key role in digestion is the pancreas. Certain *endocrine* glands secrete hormones directly into the blood; among these are the *pituitary*, at the base of the brain, the *thyroid*, in the upper chest, the two *adrenals*, one just above each kidney, and the male and female *gonads*, known respectively as the *testes* and the *ovaries*. Some aspects of the endocrine system will be discussed below, in the section on reproductive toxicity).

Excreted bile is carried by a small duct to the uppermost part of the small intestine, called the duodenum, though some is stored in a sac just under the liver called the gall bladder. Bile is joined by secretions from the pancreas, and they combine to enhance digestion; the bile has a detergent-like action on fats, breaking them up into small droplets so they can be readily attacked by digestive enzymes. Bile is heavily pigmented because it contains waste chemicals generated by the liver, including breakdown products of hemoglobin, which is collected in the liver as red blood cells age and collapse. It is ultimately eliminated in the feces, to which it imparts its characteristic colors.

The liver is a prime target for toxicity because all chemicals received orally are carried directly to the liver by the hepatic portal vein, immediately after absorption. As mentioned, liver cells have an astounding capacity to metabolize these foreign compounds, in most instances turning them into water-soluble forms that can be readily excreted from the body (through the kidney). But this detoxification capacity of the liver can sometimes be overwhelmed. Moreover, some forms of metabolic change, as illustrated in the case of bromobenzene in Chapter 3, create metabolites having toxic properties more threatening than those of the original chemical. In either of the last two conditions liver damage can occur, as can damage at other sites of the body when toxic molecules escape the liver.

Toxicologists classify hepatic toxicants according to the type of injuries they produce. Some cause accumulation of excessive and potentially dangerous amounts of *lipids* (fats). Others can kill liver cells; they cause cell *necrosis*. *Cholestasis*, which is decreased secretion of bile leading to jaundice (accumulation of gruesome looking pigments that impart a yellowish color to the skin and eyes) can be produced as side effects of several therapeutic agents. *Cirrhosis*, a chronic change characterized by the deposition of connective tissue fibers, can be brought about after chronic response to several substances. And, as will be reviewed in the next chapter, liver cells can be altered by chemicals and develop into tumors, of both benign and malignant nature. Experimentalists who study the liver's many and varied responses to chemicals will caution the reader that 'hepatotoxicity' is not a very helpful term, because it fails to convey the fact that several quite distinct types of hepatic injury can be induced by chemical exposures and that, for each, different underlying mechanisms are at work. In fact, this situation exists for all targets, not only the liver. Lipid accumulation – fatty livers – can result, for example, if a hepatotoxic chemical somehow alters biochemical pathways to produce an oversupply of the chemicals out of which fats (lipids) are synthesized. Another chemical can interfere with the process that normally breaks down liver fats, with the same result – lipid accumulation. That chemicals can cause fatty liver was first discovered more than a century ago, when worker exposure to yellow phosphorus, which was used to manufacture match heads, was found to be associated with this condition.

Carbon tetrachloride (CCl_4), once a very widely used solvent, has perhaps been the subject of more experimental study than any other organic chemical. Since the early 1920s experimentalists have been investigating its various effects on the liver and have come to understand in great detail how this molecule performs its deeds.

The carbon tetrachloride molecule has the simple chemical structure shown on the left; four atoms of chlorine are chemically bonded to one carbon atom.

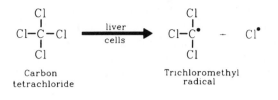

Carbon tetrachloride → Trichloromethyl radical

When molecules reach the liver some are acted upon by components of cells, which manage to break apart one of the carbon–chlorine bonds. The chemical bond consists of a pair of interacting electrons (the little dots), one contributed by chlorine, the other by carbon. The bond-breaking results in the release of an atom of chlorine and a very reactive chemical group called the trichloromethyl (CCl_3) radical, as shown in the chemical equation. It appears that damage to the liver cells is initiated by the trichloromethyl radical, which has the capacity to interact chemically with some of the normal protein and fat molecules of the cell. Those disturbances result, among other things, in the abnormal oxidation of the fats of the liver cell walls; these molecular events can result in a sequence of additional damaging events, the nature and extent of which depend upon the concentration of CCl_3 radicals generated and their lasting power within cells. Cell death – necrosis – can result if damage to cell walls is extensive. Carbon tetrachloride can act through other mechanisms and cause other types of liver cell damage. Some other chemicals, the closely-related solvent chloroform ($CHCl_3$) for example, cause liver cell necrosis in ways similar to carbon tetrachloride, while others act in quite different ways. So we may list many chemicals capable of causing specific types of liver injury, but such listing may obscure the fact that many differing underlying molecular and cellular events may be at work.

Liver cirrhosis can result from chronic exposure to several chemicals, including carbon tetrachloride, alcohol, and aflatoxin (the mold product described in the Prologue). Over time continuous liver injury leads to the accumulation of the abnormal fibers made of collagen, the normal protein component of bones. The biological mechanisms underlying these events and their progression over time are not particularly well understood. The well-known phenomenon of liver cirrhosis following long-term ingestion of excessive amounts of alcohol presents an interesting toxicological problem. Some scientists believe cirrhosis is a direct consequence of alcohol-induced toxicity.

But others believe the evidence points to a kind of indirect affect – specifically, that the cirrhosis actually results from chronic nutritional deficiency frequently associated with alcoholism.

Several clinical chemical tests are available to detect the presence of liver injury. Certain normal liver enzymes can be released to the blood following injury to the cells containing them, and a search for their presence is a routine component of chemical testing and of monitoring of animals during toxicity testing. Other tests, routinely performed during animal testing and on human beings subjected to medical examination, provide information about the nature and extent of hepatic disease.

Kidney – nephrotoxicity

The kidneys – a pair of organs in the lower back region, just below the ribs – are part of the *urinary system*, the main function of which is to rid the body of waste substances, including those resulting from normal biochemical processes and those resulting from absorption of other, non-essential chemicals. The kidneys are essentially filters, removing wastes from blood carried through them. They also play a critical role in regulating the contents of body fluids. Urine is formed in the kidneys and is carried to the *bladder* by two tubes called ureters. The urinary bladder stores urine until the volume reaches a certain level, whereupon it is released through a tube called the *urethra* to be excreted from the body.

The main filtering units of the kidneys are called *nephrons*; about one million nephrons are present in each kidney. Each nephron consists of a *renal corpuscle* and a unit called a *tubule*. Blood carrying normal metabolic wastes such as urea and creatine moves through a portion of the corpuscle called the *glomerulus*, where a filtrate forms that contains water, normal metabolic products, and also waste products; the filtrate collects in another unit called Bowman's capsule. Glomerular filtrate then moves into a highly convoluted and multifaceted set of tubes – the tubule – where most useful products (water, vitamins, some minerals, glucose, amino acids) are taken back into the blood, and from which waste products are collected as urine. The relative amounts of water and minerals secreted or returned to the blood are under hormonal control.

Chemicals toxic to the kidneys can injure different components of them and thereby undermine their functions in several ways. Several so-called 'heavy' metals, notably mercury, cadmium, chromium, and lead, are particularly ruinous to the tubules. Certain concentrations of these metals present in glomerular filtrate can seriously impair the functions of the tubules and this can lead to loss from the body of excessive amounts of essential molecules such as sugar (glucose) and amino acids. If concentrations of these metals reach high enough levels cell death can follow, which, if extensive enough, can close down kidney function altogether. The two hepatoxicants mentioned earlier, carbon tetrachloride and chloroform, also cause nephrotoxicity.

Certain antibiotics such as the tetracyclines, streptomycin, neomycin and kanamycin can cripple the tubules if taken in excessive amounts. Toxic damage to the kidneys can affect not only their filtration functions, but can alter the organs' control over blood levels of certain critical molecules. A complex biochemical–hormonal system controlling blood pressure and volume, for example, is regulated by the kidneys, so that chronic kidney damage can inflict damage on the circulatory system, including the heart itself. The kidneys also exert control over molecules critical to hemoglobin synthesis.

A number of clinical tests are available to detect kidney damage. The clinician examining a patient or the toxicologist monitoring an animal toxicity study collects urine and blood samples. Indications of kidney damage (which, of course, for the human patient could be related to many factors other then chemical toxicity) include urinary excretion of excessive amounts of proteins and glucose and excessive levels in the blood of unexcreted waste products such as urea and creatine. A number of additional kidney function tests are available to help pin down the location of kidney dysfunction.

Nervous system – neurotoxicity

The nervous system consists of two main units: the *central nervous system* (CNS), which includes the *brain* and the *spinal cord*; and the *peripheral nervous system* (PNS), which includes the body's system of nerves that control the muscles (motor function), the senses (the sensory nerves), and which are involved in other critical control functions. The individual units of the nervous system are the nerve cells, called *neurons*. Neurons are unique types of cells because they have the capacity to transmit electrical messages around the body. Messages pass from one neuron to the next in a structure called a *synapse*. Electric impulses moving along a branch of the neuron called the *axon* reach the synapse (a space between neurons) and cause the release of certain chemicals called *neurotransmitters*. These chemicals migrate to a unit of the next neuron called the dendrites, where their presence causes the build up of an electrical impulse in the second neuron.

There are three types of neurons. *Sensory* (or afferent) neurons, as their name implies, carry 'sense' information about the body; *motor* (or efferent) nerve cells carry 'instructions' from the CNS, in the form of nervous impulses, to the muscles, organs, and glands of the body. The *associated neurons*, which have several other names, are involved in detecting impulses from sensory neurons and passing them to motor neurons.

Nerves are bundles of nerve fibers, connective tissue, and blood vessels. Each fiber is part of a neuron. Afferent nerves consist entirely of sensory neurons and efferent (motor) nerves consist of motor neurons.

The nervous system has many complex components but we shall not go more deeply into its structure and functioning here. Obviously, toxic damage to certain components of it can be unfavorable for the whole organism, because the nervous system is intimately involved in control of virtually all of the body's mental and physical functions.

Two additional features of the nervous system are particularly important to a discussion of neurotoxicity. The *blood–brain barrier* and the *blood–nerve barrier* act as protective devices for the nervous system, and are effective at preventing movement of certain chemicals from the blood to the brain and nerves. Unfortunately neither barrier is effective against all types of molecules, and there are plenty of examples of brain and nervous system toxicants that can penetrate the 'barriers.'

Neurotoxicologists generally categorize effects according to the chemical's primary site of action.

Some substances, such as carbon monoxide and barbiturates, can deprive brain cells of oxygen or glucose – they produce anoxia – with potentially serious consequences for gray matter. Other substances, such as lead, hexachlorophene and the antitubercular drug isoniazid, are capable of causing loss of myelin, a coating or sheath for the axon and dendrites that extend from the central unit (cell body) of neurons. Demyelination can occur in either the CNS or PNS.

So-called *peripheral neuropathies* can result from excessive exposures to certain industrial solvents such as carbon disulfide (CS_2, used in the rubber and rayon industries) and hexane (C_6H_{14}, once used in certain glues and cleaning fluids). Over-exposure to acrylamide, an important industrial chemical, and chronic alcohol abuse can also induce this effect. As the name implies, it involves attack of the chemical on and damage to axonal portions of neurons. Typical symptoms of peripheral neuropathies include weakness or numbness in the limbs, which are more or less reversible depending upon the specific agent and the intensity of exposure.

A particularly interesting example of the role of metabolism in toxicity was brought to light during the 1970s by Herbert Schaumburg and Peter Spencer. We have mentioned that the solvent hexane (structure I) is capable of causing delayed peripheral neuropathy in the form of axonal degeneration. Excessive occupational exposure to hexane leads to loss of sensation in the limbs, particularly in the hands and feet. Muscle weakness in the extremities is common. The effects are usually delayed, and occur several months to a year following initial exposure. The active toxicant is not hexane, but the hexane metabolite called 2,5-hexanedione (II):

$$CH_3-CH_2-CH_2-CH_2-CH_2-CH_3 \qquad CH_3-\overset{\overset{\displaystyle O}{\|}}{C}-CH_2-CH_2-\overset{\overset{\displaystyle O}{\|}}{C}-CH_3$$

Hexane (I) 2,5–Hexanedione (II)

$$CH_3-\overset{\overset{\displaystyle O}{\|}}{C}-CH_2-CH_2-CH_2-CH_3$$

Methyl–*n*–butyl ketone (III)

The latter compound results from loss of two hydrogen atoms by metabolic oxidation at each of two of the carbon atoms of hexane and their replacement by oxygen atoms. Schaumburg and Spencer, and several other investigators as well, including Walter Krasavage and his associates at the Eastman-Kodak Company, not only showed this, but also showed that the identical neurotoxic events result from direct exposures to Compound II and also another common solvent called methyl-*n*-butyl ketone (III). Chemical III is, like hexane, readily metabolized to the active toxicant, molecule II. Because both I and III yield the same metabolite (II), and because this metabolite is the source of toxicity, then exposure to both of these chemicals produces the identical type of neuropathy.

Certain organic forms of mercury can elicit specific damage in the main cell body of peripheral neurons. Similar responses are associated with certain natural products called vincristine and vinblastine, both of which have been used as antileukemic medicines. The deadly botulinum toxins, mentioned in Chapter 5, block transmission of nerve impulses at the synapses of motor neurons. This blockage results in muscular paralysis which, if severe enough, can lead to death, usually because respiration is impaired. The once widely-used pesticide, DDT, is an organic chemical that also acts on the nervous system at this site, though it also can mount an attack on areas of the CNS.

The CNS is a target for many neurotoxicants, including mercury, certain forms of gold, manganese, and even the food ingredient monosodium glutamate (MSG), although very large doses of the latter are required. Depending upon their location and severity, effects on the CNS can have profound consequences for sensory and motor functions. Chronic manganese intoxication, for example, produces signs and symptoms that in advanced stages mimic the condition known as parkinsonism. The most common early indications of chronic poisoning with this metal include incoordination, difficulty in keeping a steady gait, impaired speech, and severe tiredness. Tremors, weakness in the limbs leading to an inability to stand upright, and even emotional disturbances appear at more advanced stages. Severe tremors during resting is a particularly telling characteristic of manganese poisoning. Removal of the worker from manganese environments and perhaps treatment with agents that assist removal of manganese from the body are necessary to alleviate this severe condition.

Excessive exposure to inorganic mercury, particularly in its elemental form, creates a psychological condition called erethism. Victims suffer from excessive timidity and self-consciousness, inability to concentrate, loss of memory, and other psychological changes. From at least the 17th and well into the 19th century, mercury was used to cure felt, and workers exposed during that process could acquire erethism. Lewis Carroll's character the Mad Hatter was based no doubt on the fact that hatters exposed to mercury could in fact go mad. The phrase 'mad as a hatter' was in common use at the time *Alice's Adventures in Wonderland* was written.

A rather insidious form of mercury neurotoxicity can occur when the element is bound chemically to certain groupings of organic molecules – so called alkylmercury compounds. Such forms of mercury, unlike the inorganic forms, can cross the blood–brain barrier (which is fairly effective at excluding inorganic mercury) and damage brain cells in serious ways. The most devastating and best-studied manifestation of this form of neurotoxicity occurred at Minamata, Japan, during the 1950s and 1960s, and perhaps continued to the mid-1970s. Discharge from a chemical plant of elemental mercury into Minamata Bay created high levels of the element in the water. When in this environment, the mercury can be converted to the organic form by microorganisms present in sediments. The organic mercury compound found its way into fish and shellfish, and thence to humans, mostly local fishermen and their families. The disease was first identified as organic mercury poisoning in 1963. The patients exhibited visual and hearing impairments, and mental disturbances characterized by periods of agitation alternating with periods of stupor. Similar outbreaks of organic mercury poisoning occurred in Iraq in 1956, 1960, and 1971; in these cases the source was grain that had been treated with an organic mercury compound to prevent mold growth. The grain was supposed to be used only for seed, but was instead taken for grinding to flour. The 1971 outbreak affected more than 6500 people, of whom more than 400 died.

Perhaps the most important of the neurotoxic agents found in the environment is lead (chemical symbol, Pb, from the Latin word for lead, plumbum).

People have been extracting lead from ores for about 4000 years. Its uses have been many and have changed over the centuries, sometimes because of concerns about the metal's dangers to health. Lead has probably been the subject of more toxicological investigations than any other substance. An unusually high proportion of these investigations involve observations in humans. There are two reasons for this: (1) human exposure to this metal has been and still is very widespread; and (2) a means has been available for some time to obtain a direct measure of that exposure. The latter is the concentration of lead in human blood, commonly expressed as micrograms Pb per deciliter (100 ml) of blood (μg/dl). It is relatively easy to obtain this value by taking a sample of blood and subjecting it to chemical analysis.

Young children are especially susceptible to the effects of environmental lead, first because their bodies accumulate lead more readily than do those of adults and, second, because they appear to be more vulnerable to certain of the biological effects of lead. In 1988 the U.S. Public Health Service estimated that, in the United States alone, 12 million children are exposed to leaded paint, 5.6 million to leaded gasoline, 5.9–11 million to dusts and soils containing excessive lead,

10.4 million to lead in water (in part because of lead in pipe solders) and 1.0 million to lead in food. (These figures should not be added, because some children are simultaneously exposed to more than one of these sources.) The Public Health Service also estimated that nearly 2.4 million children in the United States exhibit blood lead values greater than 15 μg/dl (the upper end of the range now considered tolerable by the Public Health Service), and that nearly 200 000 children exhibit levels greater than that at which some form of medical intervention is appropriate. Although substantial progress in reducing lead body burdens has been made in the past two decades (e.g., by programs to eliminate leaded gasoline and to prevent children's exposure to leaded paints present in many older homes), much remains to be done. The problem is complicated by the fact that as we have worked to reduce lead exposures, health scientists have uncovered new concerns.

Lead has multiple toxic effects, on elements of the blood, where it interferes with heme synthesis, and on the kidney. But of greatest concern are lead's effects on the central nervous system. Encephalopathy – brain disease – and peripheral neuropathy occur when blood levels reach the range of 70–100 μg/dl, a level which is nowadays rare in children, at least in the United States. Lead encephalopathy manifests itself as a stupor, coma, and convulsions, and is generally accompanied by severe cerebral edema and other types of brain cell damage. Until fairly recently a blood lead level of 25 μg/dl or less was thought to be without much medical significance and was sufficiently low to prevent encephalopathy, peripheral neuropathy and even other, less serious manifestations of neurologic injury. But beginning in 1979, when Herbert L. Needleman and his associates at the University of Pittsburgh School of Medicine reported so-called 'subclinical effects' at blood lead levels below 25 μg/dl, and continuing over the past decade as Needleman, and other investigators as well, uncovered such effects at increasingly lower blood lead levels, the protective value of the 25 μg/dl limit has come increasingly to be questioned.

What are these 'subclinical effects'? Very simply, they are effects that occur at blood lead levels below those that produce clinically measurable effects – they occur in the absence of any sign of overt lead poisoning. These effects can be detected only by studying various forms of behavior, such as degree of hyperactivity and classroom attention span, and performance on various tests of intelligence and mental development. Deficits in neurobehavioral development, as measured by two widely used tests – the Bayley and McCarthy Scales – have been reported in children exposed prenatally (via maternal blood) to blood levels in the range of 10–15 μg/dl. Small IQ deficits have also been reported at blood lead levels of 25 μg/dl. The extent to which these various deficits persist beyond the first several years of life is not known. Indeed, some investigators have not been able to measure such deficits, at least at these low blood levels, in the populations they have studied, so the issue remains somewhat controversial.

Needleman and others who are studying this problem observe small shifts in, among other things, the distribution of IQs in populations of children exhibiting blood lead levels in a range below that known to produce clinical effects. Are such shifts of medical or social importance? For the individual child a small downward shift in IQ, for example, is probably not of significance, especially if it does not persist beyond early childhood. But for society as a whole such a shift may be of substantial significance if millions of children are included, and the effect persists into later life. This is a considerable dilemma that public health officials have not yet fully faced, although it is clear the trend in public health policy is to move toward the goal of reducing blood lead levels.

Signs of neurotoxicity can be detected in several ways. Assessment of mental state and sensory functions can be performed in humans; in fact it is difficult or impossible to make such assessments in experimental animals (you don't get much of a response from a rat when you ask if he's often feeling tired, has recurring headaches, or is having difficulty seeing). There are objective tests of sensory function, particularly as it involves the optic and auditory nerves; responses to light and sound stimuli can be readily measured. Motor examination includes inspection of muscles for weakness and other signs of dysfunction, such as tremors. Specific types of motor disorders give an indication of the part of the nervous system that has been affected.

Reflex action can be objectively examined. Measurement of the so-called electrical conduction velocity of motor nerves gives an indication of whether peripheral nerves have been damaged. Electromyography, or examination of the electrical activities of the muscle, and electroencephalography (EEG) can also be used to detect the presence of neurologic abnormalities. And, at least in animal tests, detailed pathological examinations of neural tissue can be performed on animals that are killed.

A particularly rapidly growing area of the science is *behavioral toxicology*. The nervous system serves a vast integrative function for the body, and various forms of behavior can be altered by responses to substances that affect the nervous system in some way. We have already seen some examples of behavioral changes associated with exposures to lead, mercury, and manganese. The incorporation of behavioral measures into animal tests is increasingly common, but some aspects are controversial. Does, for example, evidence of decreased body movement (objectively measured) in chemically-treated mice signal that the causative agent poses some type of hazard for humans? The relevance to human health of observations such as these is not quite as straightforward as, say, the observation that a chemical demyelinates nerve fibers. Nevertheless the study of the influences of chemicals on human behavior, psychological health, and intelligence, is a dynamic area of research, and is likely to occupy a significant position in the total evaluation of chemical risks in the not-too-distant future.

Reproductive system toxicity

The reproductive systems of both males and females can be harmed by certain chemicals. In males certain chemicals cause the testes to atrophy and reduce or eliminate their capacity to produce sperm. Particularly striking in this regard is a now-banned but once widely-used pesticide called DBCP, residues of which persist in ground water supplies in a few regions of the country. Its pronounced impact on spermatogenesis is readily detectable in experimental animals and, unfortunately, has also been observed in some men once occupationally exposed to large amounts. The heavy metal cadmium is another substance effective at reducing sperm production.

Impairment of the reproductive process can occur in several other ways, generally by the induction of abnormal physiological and biochemical changes that reduce fertility, prevent the full maturation of the fetus, or prevent successful birth (parturition). Reproductive success, or lack thereof, is measured in animal experiments in which mating is allowed to occur between chemically-treated females and untreated males, or vice versa, or between males and females that are both exposed to the chemical. Various indices of reproductive performance are then measured. The fertility index, for example, is the percentage of matings that result in pregnancies. The gestation index is the percentage of pregnancies resulting in the birth of live litters. The percentage of offspring that survive four days or longer is called the viability index. These and several other indices are the measures by which treated and control animals are compared.

A particularly well-studied area of reproductive toxicity is called *developmental toxicity*. Here we are concerned about the effect of the chemical on the developing embryo and fetus (exposures received *in utero*), and on the further development of the infant and child subsequent to birth, at which time chemical exposure may cease, or may continue to weaning because the chemical received by the mother is transferred to her milk, or may continue throughout life because there are sources of the chemical in addition to that supplied by the mother because of her exposure. We saw in the section on neurotoxicity how lead exposures *in utero* may impair development following birth. Of course, some chemicals are severely toxic to the embryo and fetus and can cause their deaths prior to birth. A particularly important area of developmental toxicity is called *teratogenicity*, treated in the next section.

Teratogenicity

After fertilization the ovum – a single cell – begins to proliferate, making more of its own kind by a series of divisions. In humans, at about the ninth day the remarkable process of cell differentiation begins; the specific types of cells (e.g., neurons, liver cells, etc.) that make up the body begin to form and to migrate to their appropriate positions. This *embryonic* period lasts until the 14th week of gestation in humans. This is the period of *organogenesis*, which means exactly what it appears to mean – the various organs of the body are generated, although at different times and rates. Following the embryonic is the *fetal* period, during which the organism grows and bodily functions mature. The fetal stage ends at birth, although obviously development continues to take place following this event. Although the total duration of embryonic and fetal development varies among species, the sequence and relative timing of the critical events is about the same in all mammals.

Thalidomide, the chemical structure of which is shown, was introduced in 1956 by a German pharmaceutical firm for use as a sedative. The drug was widely used, though not in the United States, at oral doses of 50–200 mg/day to reduce nausea and vomiting during pregnancy and it was quite effective at doing so. Few side effects were experienced by women taking the drug.

An unexpected increase in the incidence of certain rare and devastating birth defects – absence of limbs and reduced limb length – was reported beginning in 1960, first in West Germany and then in other areas of the world. Thalidomide was identified as the cause through the

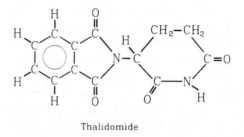

Thalidomide

work of W.G. McBride in Australia and W. Lenz in West Germany. The drug was taken off the market in 1961 and the increased incidence of these rare birth defects disappeared at the predicted time thereafter. Very few cases were reported in the United States, where full scale entry into the marketplace was delayed because of some careful work by Dr. Frances Kelsey of the FDA. Perhaps 8000–10 000 cases were reported worldwide. The offspring of women taking the drug during the sixth and seventh weeks of pregnancy were found to be at highest risk of developing birth defects. This is the period of organogenesis in which the skeleton is under most rapid development.

Thalidomide is a *teratogen*, in humans and in experimental animals, although the latter were found susceptible to these effects only after an extensive set of investigations was undertaken following the tragic human findings. 'Teras' is the Greek word for 'monsters'. That some humans are born with structural abnormalities of the body – defects of the palate, of the skeletal system, of the heart, the eye, and so on – has been known since antiquity, and the appearance of human 'monsters' of one sort or another in legend, myth, and in actuality is a prominent feature of human history. Clearly teratogens and other factors that cause birth defects have been around for a very long time and predate the chemical revolution by at least several millennia.

The more common medical term for birth defects is 'congenital abnormalities,' meaning those that are inherited. The earliest medical hypotheses suggested a genetic cause for these abnormalities, not an environmental one. This is probably still true for many such abnormalities, but in the 1930s it was learned that terata could be induced in animal species by manipulating their environments – in this case by withholding adequate amounts of riboflavin or Vitamin A from pregnant females. Since that time several environmental agents have been found capable of causing 'congenital' abnormalities in humans, and these and a great many more have been found to be animal teratogens. It is also noteworthy that livestock and other domesticated animals as well as wild animals are all susceptible to environmental teratogens, and some of the most potent teratogens occur as natural components of certain range weeds that are sometimes consumed by cattle or sheep.

Experimental studies reveal quite conclusively that the timing of exposure to a teratogen – the time during gestation the dose is received – is as important as the size of the dose. Because the rates and timing of cell differentiation and organogenesis vary among organs, they will be differently affected depending upon the period of dosing. In the rat, for example, dosing early in the period of organogenesis tends to cause terata of the eye and brain much more frequently than defects of the skeleton or the urinary system or the palate, while the opposite is true for dosing during the mid-point of that period. A few days can make a large difference in the severity produced by and target for a teratogen. (Note that, with teratogens, we are no longer grouping toxicity by target. Specific teratogens do have specific organ or system targets, but chemicals in this class are grouped together because they produce birth defects.)

Typical animal tests for teratogens involve rats and rabbits, and sometimes other species. The agent to be tested is administered to groups of pregnant animals at several dose levels during the entire period of organogenesis (days 6–15 of gestation in the rat). Fetuses are usually removed from the mother one day prior to expected delivery, and various examinations are made to determine the fate and health status of each fetus and to identify physical abnormalities. Because dosing occurs throughout organogenesis these tests can capture a broad range of possible teratogenic effects.

A particularly important aspect of the teratogenicity study concerns the effect of the chemical on the treated mothers. If a dose that is in some way toxic to the mother is used in these tests – and the highest dose is often toxic in this way – the test results may not be telling with respect to the offspring. Thus, if the offspring from mothers that experience a toxic response are also affected, it will not be clear whether the chemical is itself teratogenic, or whether the health of the fetus was injured only because of its mother's unhealthy condition. Generally, toxicologists do not consider an agent teratogenic if its effects on offspring appear only at the 'maternally toxic dose,' and not at lower doses.

In keeping with the earlier discussion of developmental toxicity we note that toxicologists nowadays take a broad view of teratogenicity;

they consider not only *structural* but also *functional* abnormalities to qualify as birth defects, as long as they were produced as a result of exposures incurred *in utero*. Thus, for example, the developmental effects of chronic alcohol abuse by pregnant women, categorized as Fetal Alcohol Syndrome, are characterized not only by the presence of certain craniofacial abnormalities, but also by a variety of disabilities such as shortened attention span, speech disorders, and restlessness. Although fully-expressed physical deformities included in FAS are associated with heavy drinking, debate continues on the level of alcohol consumption, if any, that is without these more subtle effects on behavior.

Another feature of developmental toxicity is raised by the experience we have had with the drug DES – diethylstilbestrol. DES is an easily synthesized chemical having potent *estrogenic* properties – its chemical characteristics are such that it can fulfill many of the biological functions of the normal female sex hormones, or estrogens. The drug was widely used in pregnant women from the mid-1940s until 1970 to prevent threatened miscarriage. In the late 1960s, Arthur Herbst of the Massachusetts General Hospital reported an unusual occurrence of a cancerous condition called clear-cell adenocarcinoma of the vagina. What made the occurrence unusual was the fact that the seven cases Herbst saw in the 1966–1969 period were in women between the ages of 15 and 22. This particular cancer was extremely rare in this age group.

A series of investigations initiated during this period led to the conclusion that DES use by the mothers of these young women was the causative factor. DES is a teratogen, because its effect was the result of exposures received *in utero*, but that effect – a rare form of vaginal cancer – was not fully expressed for about two decades following the exposure (the age at which the incidence of the disease peaked was later shown to be 19). DES produced not only cancer but abnormalities of the sexual organs in both male and female offspring, most of which were not expressed until the children passed the age of puberty.

It should not be assumed that all birth defects and functional abnormalities in children are caused by drugs or environmental chemicals. It is clear that environmental factors such as extreme heat or cold, certain forms of radiation, infections (particularly German measles and syphilis), dietary deficiencies, and genetic disorders in the parents can all put the developing fetus at risk.

Because our intention in this book is to emphasize principles and not toxic agents, we shall here complete the survey of slow poisons and their targets. Although many of the examples presented are environmental chemicals of considerable importance (e.g., CO, Pb, Hg), the examples were chosen primarily to illustrate certain principles of toxicity: the importance of dose and exposure duration, of the chemical form of the toxic agent, of exposure route, and especially in the case of reproductive toxicants, of the timing of exposure. They also illustrate the role of ADME in the production of toxicity and the fact that toxicity targets are many and varied; in fact, we have seen that chemicals exert their effects in many different ways on the same target – not all substances listed as liver toxicants act in the same way. Our survey is, however, highly incomplete and by no means reflects all that toxicologists know about 'slow poisoning'. It nevertheless should serve to set the stage for the later discussions of dose–response and the ultimate issue of human risk.

7

Carcinogens

People all over the world are exposed to cancer-causing chemicals present in air, water, food, consumer products, and even in soils and dusts. There are probably a few dozen known chemical carcinogens reaching people through these media and, as a rough guess, a few hundred more yet to be identified. In their places of work some people come into contact with additional cancer-causing agents, generally at higher exposure levels than those experienced by the general population. Some people deliberately expose themselves, and incidentally expose others, to the large number of known and suspected carcinogens present in tobacco smoke. People also get exposed to various physical agents – ultraviolet radiation from the sun and sunlamps and other forms of natural and man-produced radiation – that increase cancer risks. We are all being assaulted by chemical and physical carcinogens. No wonder the chance of developing some form of cancer over our lifetimes is about one in five!

But we are moving too quickly. Before we can begin to contemplate the contribution of all these environmental carcinogens to the total cancer problem we need to acquire a better understanding of what is meant by the terms 'carcinogen' or 'cancer-causing chemical' and of how certain substances get to carry those labels. Carcinogenicity is one form of slow poisoning, but an especially complex one, in both scientific and social terms, so it deserves its own chapter.

We shall begin with a little history, and then move to a discussion of how carcinogens are identified. Here, for the first time, the methods and limitations of epidemiological science are given extensive treatment, because it has made especially important contributions to identifying cancer-causing substances. Animal tests used to identify carcinogens, while similar in many respects to those reviewed in the last chapter, have some special features that need to be brought out. Here we shall also compare results from animal studies and epidemiological investigations.

Cancer and chemical carcinogens

Professor Michael Shimkin, a leading figure in cancer research for many decades, has written the following description of this greatly feared disease:

First of all, cancer is a word in the English language, derived from the Greek word for crab, *karinos*.[9] Among its many synonyms are malignant tumor and malignant neoplasm (from the Greek for new growth). Subgroups of cancer, describing the body tissues of origin, include carcinoma, sarcoma, melanoma, lymphoma, and many other related or combined terms.

Cancer is a word that stands for a great group of diseases that affect man and animals. Cancer can arise in any organ or tissue of which the body is composed. Its main characteristics include an abnormal, seemingly unrestricted growth of body cells, with the resulting mass compressing, invading and destroying contiguous normal tissues. Cancer cells then break off or leave the original mass and are carried by the blood or lymph to distant sites of the body. There they set up secondary colonies, or metastases, further invading and destroying the organs that are involved.

Another important characteristic of cancer is its appearance under the microscope. The individual cells vary in size and shape, and the orderly orientation of normal cells is replaced by disorganization that may be so complete that no recognizable structures remain.

At the present time, cancers are classified by their appearance under the microscope, and by the site of the body from which they arise. By such criteria of appearance (morphology) and localization, devised by pathologists during the past century, at least 100 different cancers are identified. The number can be increased to 200 or more if finer details of morphology are taken into consideration.

And . . .

For convenience, we have discussed cancer and normal cells as two sharply separate, different entities. Actually, there is a wide spectrum of cellular abnormalities and normal reactions between these extremes. The rate of cell

[9] The image of malignancies as crabs appears first to have occurred to Galen, the great Greek physician who practiced in Rome in the second century A.D.

division and tissue growth of a normal pregnancy exceeds the rate of growth of many cancers, with the key difference being that the normal process stops when it has reached its end point. In appearance also, the normal repair of injured tissue may for a time look quite 'wild' under the microscope, but represents a self-limiting reaction. There is also a spectrum between the benign, or innocent tumors, and cancers, including a zone of borderline lesions regarding which decision is particularly difficult. As a further complication, a tumor that may seem benign in microscopic appearance can be fatal if located in a vital area or in a limited space, such as the brain.

It is certain that these diseases are not totally products of the industrial age or the era of modern chemical technology. Lack of solid statistical information forces us to avoid the question of how much human cancer there would now be if the industrial revolution and its chemical and physical products had never appeared on earth, but if the average age at death had nevertheless increased exactly as it has over the past two centuries. This is, of course, an unlikely historical scenario, in that products of the industrial revolution have contributed substantially to the fact that more people are living to old age. Particularly important have been medicines that prevent deaths in early childhood, antibiotics that cure infections, agricultural technology, and many forms of medical technology and public sanitation. We also bring up the issue of average age at death because most cancers are diseases of old age. If many people die early from other diseases, then the numbers of people alive to contract cancer are fewer. To be meaningful, statistics on cancer rates must contain an adjustment for differences in the distribution of ages in the populations under investigation, whether comparisons are being made for the same population at different points in time, or for different populations at the same point in time.

Human cancers were much discussed by Galen and most medical commentators ever since, and dozens of hypotheses regarding the origins (*etiologies*) of these diseases are recorded in the medical literature. A seminal event relevant to our present concerns about the environment occurred in 1775. A British surgeon, Percival Pott, published his observations on high rates of cancers of the scrotum among London chimney sweeps. Pott attributed the cancers to the soots with which these workers came into contact. The surgeon reached conclusions about the causal relationship between soots and scrotal cancers on several grounds, not least of which was the fact that the occurrence of these cancers could be reduced if certain hygienic practices were followed to reduce the sweeps' direct contact with soots.

Pott's observations can be said to be the first to establish reliably a cause–effect relation between an environmental agent and cancer, and also to recognize the importance of good industrial hygiene measures to protect workers from hazardous agents.

Pott knew nothing about the chemical composition of soot. We now know that soots are mostly composed of inorganic carbon (a biologically rather inert material which is also the major ingredient of charcoal and, in crystalline form, of diamonds), but they also contain small amounts of many different chemicals that are grouped under the general heading of polycyclic aromatic hydrocarbons (PAHs). PAHs occur as degradation products whenever any organic materials – fuels, foods, tobacco, for example – are burned or heated to high temperatures. These chemicals are also present in unburned petroleum and products such as coal tars. Occupational skin cancers associated with materials related to soots were reported by several investigators in England and Scotland during the last quarter of the nineteenth century. Ross and Cropper, two British scientists, proposed in 1912 that coal-tar related cancers were induced by chemicals, the same chemicals also found in soots and combustion products of various sorts.

Although they did not know it at the time, two Japanese scientists, Katsusaburo Yamagiwa and Koichi Ichikawa, provided in 1915 indirect experimental confirmation of Pott's observations on soots when they were able to produce skin tumors on the ears of rabbits to which they had applied coal tar (not soot) for many months. The work of the Japanese investigators is also significant because it represented the first laboratory production of tumors with an environmental chemical, or chemical mixture. (Certain animal cancers had already been shown to be produced by viruses by Ellerman and Bang (1908) and Rous (1911).) Of particular interest was their observation that tumors appeared only after many months of continuous application of the cancer-causing agent. That studies of chronic duration are necessary to detect most carcinogens has been amply confirmed since the pioneering work of Ichikawa and Yamagiwa.

In the 1920s two British scientists, E.L. Kennaway and I. Heiger, suspecting that the carcinogenically active components of coal tar were to be found among the PAHs, tested one member of the class called dibenz[a,h]anthracene on the skin of shaved mice and found it to be carcinogenic. This work, reported in the *British Medical Journal* in 1930, was the first in which a single chemical compound was shown to be capable of producing tumors. A team of chemists at the London Free Cancer Hospital processed about two tons of coal tar pitch and isolated small amounts of a pair of isomeric PAHs, benzo(a)pyrene and benzo(e)pyrene. The former proved to be carcinogenic, the latter not. The chemical structures of these two PAHs and Kennaway and Hieger's PAH are as shown, in a shorthand form:

Dibenz[a,h]anthracene

Benzo[e]pyrene

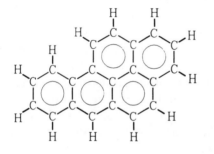

Benzo(a)pyrene
(long form)

Benzo(a)pyrene
(abbreviated form)

These chemical structures are simply abbreviated forms of those introduced in Chapter 1. PAHs are composed of carbon and hydrogen atoms only. The form of the abbreviation is illustrated with benzo(a)-pyrene, which is shown in both abbreviated (on the right) and 'long' forms (on the left). In the 'long' form all carbon and hydrogen atoms are explicitly shown, with each carbon atom carrying the required four bonds. Note that the carbon atoms are arranged in rings of six each – they are 'cycles' of carbon atoms, all of which contain six electrons (represented by the circle). Because PAHs contain several of these rings of carbon fused together, they are called 'polycyclic.' They are 'hydro-carbons' because they contain only carbon and hydrogen atoms; they are 'aromatic' for reasons having to do with the nature of the chemical bonding in each of the rings (although the term originally was used in its popular meaning, it has assumed a technical meaning having nothing to do with odors), represented by the circle that depicts a set of six electrons. The abbreviated chemical structure does not explicitly show the 'C' and 'H' symbols, but rather their presence is implied: carbon atoms are at each of the six corners of the hexagons and hydrogen atoms are assumed to be present wherever necessary to complete the four bonds that each carbon atom must carry (compare the two forms of benzo(a)pyrene, atom for atom).

Many PAHs are present in smoke and other products of combustion, and in pitches and tars. Some are carcinogens, others are not; carcinogenicity strongly depends upon details of chemical structure, specifically the ways in which the carbon rings are attached to each other. PAHs are an important class of environmental pollutants, because of their widespread occurrence.

Innovations in chemical synthesis of dyes, as mentioned in Chapter 1, gave rise to one of the first major chemical industries. Following up on the work of the German physician Ludwig Rehn, who reported large 'clusters' of bladder cancer cases among dye workers in the 1890s, occupational physicians began during the 1930s to study systematically the persisting high rate of this disease among dye workers. A decade or more of research by epidemiologists, occupational physicians, and chemists led to the identification of a number of substances called *aromatic amines* or *amino-azo* compounds as the culprits. The work of people such as Wilhelm Hueper on bladder cancers in the dye industry provided a major impetus to research and testing to identify other chemical carcinogens to which workers and the general public might become exposed. In 1937, Hueper and his associates at the National Cancer Institute (NCI) reported the experimental production of bladder tumors in dogs, from administration of the aromatic amine called 2-naphthylamine (see structure).

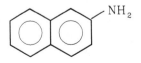

2–Naphthylamine

Hueper and several colleagues at the NCI were instrumental in drawing public attention to the issue of carcinogens in the workplace and the general environment during the two decades following the work on bladder cancer. Hueper's work and opinions were favorably cited many times by Rachel Carson in *Silent Spring*. Ms. Carson wrote:

Among the most eminent men in cancer research are many others who share Dr. Hueper's belief that malignant diseases can be reduced significantly by determined efforts to identify the environmental causes and to eliminate them or reduce their impact.

Beginning in the late 1940s, when Dr. Hueper established an Environmental Cancer Section at the NCI, and continuing to this day, a major program of carcinogen identification using animal tests has been conducted at the National Institutes of Health (NIH) (although at the present time most of this testing is conducted under the auspices of a government-wide activity called the National Toxicology Program (NTP) centered at the National Institutes of Environmental Health Sciences in Research Triangle Park, North Carolina). Several hundred chemicals have been tested for carcinogenicity by the NIH/NTP. NIH/NTP and other federal health and regulatory agencies also sponsor epidemiology studies and other investigations into the underlying chemical and biological mechanisms by which some chemicals transform normal cells to malignant ones. In addition to government-supported work, there is substantial industrially-sponsored testing and research of the same type, some of it performed because of regulatory requirements. The field of chemical carcinogenesis is a vast scientific enterprise, not only in the United States, but throughout the world.

One important part of this vast enterprise is the International Agency for Research on Cancer (IARC), a part of the World Health Organization, headquartered in Lyon, France. One of IARC's many activities involves convening meetings of scientific experts from throughout the world to examine published scientific work relating to the carcinogenicity of various chemicals. The IARC periodically publishes the results of the deliberations of these working groups. The agency also categorizes chemicals based on the nature and extent of available scientific evidence concerning their carcinogenic activity. Evidence is labelled as 'sufficient,' 'limited,' or 'inadequate,' and the reviewed chemicals are grouped into these categories; distinctions are made between evidence based on studies in human populations (case reports and epidemiology studies) and that based on studies in laboratory animals.

Table 4. *Some of the 39 Chemicals and Occupational Exposures listed by IARC as Carcinogenic to Humans.*

Note that in many cases data on cancer rates were collected under exposure conditions that no longer exist.

Some Occupational Exposures
   Boot and shoe manufacture (certain exposures)
   Furniture manufacture (wood dusts)
   Nickel refining
   Rubber industry (certain occupations)
   Underground hematite mining, when radon exposure exists.

Some Chemicals
   Arsenic and arsenic compounds
   Asbestos (when inhaled)
   Chromium and certain chromium compounds (when inhaled)
   Benzene
   Diethylstilbestrol (DES)
   2-Napthylamine, benzidine (starting materials for manufacture of certain
      dyes)
   Vinyl chloride (starting material for PVC plastic manufacture)
   Mustard gas

Some Chemical Mixtures
   Tobacco smoke
   Smokeless tobacco products
   Soots, tars, mineral oils*
   Analgesic mixtures containing phenacetin

---

* Mineral oils now in commercial production generally do not have the PAH content they had at the time the evidence of carcinogenicity was gathered.

In Table 4 are presented some of the chemicals and occupational settings IARC has categorized as carcinogenic to humans – the data from all the available epidemiology studies and case reports were considered sufficient by the expert groups to conclude that a *causal* relationship exists between exposure to the chemical or occupational setting and some form of human cancer.

Several comments should be made on the contents of Table 4. First, it is not complete; IARC now lists a total of 39 chemicals and chemical processes as carcinogenic to humans; the agency lists another two dozen or so as having limited evidence of human carcinogenicity.

Second, occupational exposures are listed instead of individual chemicals for those cases in which several chemicals may be involved in the exposure and it is not clear which one(s) is responsible for the cancer excesses observed in the workers studied. Third, the listed chemicals and occupational exposures have been demonstrated to be human carcinogens only for the specific groups of individuals upon whom observations were made by epidemiologists; whether they increase cancer risk for other individuals exposed under quite different conditions is a separate question, to be treated later in this chapter and in the chapter on risk assessment. Finally, it should be emphasized that there are many additional animal carcinogens listed by IARC and NIH/ NTP that may also pose a carcinogenic risk to humans, but for which sufficient epidemiology data have not been collected; keep in mind that a chemical can be demonstrated to be a human carcinogen only if an opportunity exists to study it in exposed humans in a systematic way, and such opportunities are not frequently found.

We now need to inquire about the scientific bases for identifying human and animal carcinogens to understand how these substances come to be present on lists such as that in Table 4. But before looking into the innards of the epidemiological and experimental methods for identifying carcinogens, it will be useful to say something about the importance of the environmental chemicals we are emphasizing in this book relative to other factors known or suspected to be involved in human cancer production.

Causes of human cancer

With a few exceptions, cancer experts generally can not determine with high confidence the specific cause of cancer in an individual. At best they can understand the factors that contribute to the cancer rates observed in large populations. Differences in the rates of certain types of cancers in different regions of a country, different countries of the world, and in the same population studied at different times, provide some indication of the relative importance of various factors. Epidemiologists also learn a great deal from studies of specific exposure situations. Several trends emerge from these types of investigations:

(1) Somewhere between 70% and 90% of human cancers appear to be of environmental origin. Here 'environmental' is used very broadly, and refers to anything not genetic. It refers not only to industrial chemicals and

pollutants, but includes factors such as diet, sexual habits, smoking behavior, and natural and manmade radiation.

(2) Many cancers are not caused by individual carcinogenic factors, but by several factors. This view is consistent with our understanding that the transformation of a normal cell to a malignant one occurs in steps, and that different agents may be involved at different steps (see the later chapter on Mechanisms).

(3) In many cases a single factor may be so important that it is considered 'the cause'. Cigarette smoking, for example, is an important cause of lung cancer because in the absence of this habit about 85% of lung cancers in males would be avoided.

(4) It has become customary among cancer epidemiologists to talk about certain 'lifestyle' factors as important contributors to cancer risk. Lifestyle factors are primarily dietary, reproductive, and other habits that are assumed to be largely under the control of individuals. These are distinguishable from factors that are less directly in the control of individuals (occupations, medicines, consumer products), and those over which individuals have little or no control (food additives, pesticides, environmental pollutants).

In 1981, two eminent British cancer experts, Sir Richard Doll and Richard Peto published a paper in the *Journal of the National Cancer Institute* entitled 'The causes of cancer: Quantitative estimates of avoidable risks of cancer in the United States today.' The authors drew upon a vast body of literature of the type mentioned above, and attempted to allocate the deaths caused by cancers among various responsible factors. The authors concluded that a certain per cent of human cancer deaths could be avoided if exposure to the responsible factors could be eliminated or controlled in some way, although the appropriate degree and nature of control for some of the 'lifestyle' factors, especially diet, is still highly uncertain. The Doll and Peto estimates are presented in Table 5. The factors are listed in a somewhat different order than they were listed by the original authors, because of our interest in clearly separating 'lifestyle factors' (the first four listed) from those that are the more direct subject of this book; this change in no way distorts the original authors' conclusions.

What is striking about the Doll–Peto estimates is the relatively small fraction, perhaps 5–8% of human cancer deaths in the United States, that are attributable to industrial chemicals present in the work place, food, medicines and involved in environmental pollution! If these estimates are correct, then no more than about 25 000–40 000 out of the nearly half-million annual cancer deaths in the United States are

Table 5. *Proportion of avoidable human cancer deaths for both sexes of the United States population*
R. Doll and R. Peto, 1981. *Journal of the National Cancer Institute*

Factor	Per cent of total cancer deaths	
	Best estimate	Range
Tobacco	30	25–40
Alcohol	3	2–4
Diet	35	10–70
Reproductive and sexual behavior	7	1–13
Occupation	4	2–8
Food additives	less than 1	minus 5–2
Pollution	2	less than 1–5
Industrial products	less than 1	less than 1–2
Sunlight, UV-light, other radiation	3	2–4
Medicines, medical procedures	1	0.5–3
TOTAL	85–87	

Note: The remaining 13–15% are due to infectious agents (certain viruses and parasites) and certain genetic factors that predispose certain individuals. The authors have recently reduced the per cent associated with occupation to about 1. The 'minus' end of the range for food additives takes into account the fact that some of these substances, particularly the antioxidants, may protect against certain cancers.

principally the result of the industrial products that are the main subjects of regulatory interest. While this number is hardly trivial, it pales in comparison with the more than 400 000 annual cancer deaths and nearly one million new cases diagnosed each year that might be avoided if we could get people to stop smoking, drinking and sunning to excess, and to eat the right diet (although, except for increased fiber consumption and reduced fat intake, and increased fruit and vegetable consumption, we are not sure what dietary regimen is optimum to avoid these cancers and whether such a regimen might not put people at increased risk for other diseases – dietary issues are extremely complex).

So why so much attention to what most people think of when they think of carcinogens – i.e., environmental chemicals of industrial origin? First, it ought to be made clear that while the Doll–Peto estimates are acknowledged by many cancer experts as close to the mark, and are in rough agreement with estimates made by others, they are still uncertain and have been criticized by some experts as possibly misleading. Other experts argue, for example, that since cancer can take several decades to develop, the full effect of the massive increase in industrial chemical production, usage, and waste disposal that occurred following World War II is not reflected in the cancer rate statistics relied upon by Doll and Peto, which were collected primarily in the 1970s. We need to wait at least until cancer rate statistics for the 1990s have been thoroughly collected before we can know whether the chemical revolution has had a major impact on cancer incidence or death rates.

A 1990 publication in the British journal *The Lancet*, jointly authored by scientists from the United States, England, and the World Health Organization, contains an evaluation of more recent cancer mortality trends in five highly industrialized countries: France, West Germany, Italy, Japan, England and Wales, and the United States. These authors report a shift over the past two decades in certain patterns of cancer mortality. Specifically, they note that certain cancers – brain and other CNS cancers, breast cancer, multiple myeloma, kidney cancer, and non-Hodgkin lymphoma – have increased in both males and females, aged 55 and older, during this time in at least these countries. Stomach cancer continues to decline, as it has since the 1930s. It is too early to grasp the meaning of trends such as these, but such data suggest it would be premature to offer a full evaluation of the effects of the enormous increase in chemical production of the past four to five decades. The Doll–Peto estimates, and others as well, nevertheless suggest a primary role for 'lifestyle factors,' not chemical pollution and industrial products, in cancer causation.

A second issue that apparently elevates public anxiety about industrial products to a high level concerns the fact that people feel they have little or no personal control over those products. Tobacco and alcohol usage, diet, reproductive and sexual habits, and sunlight exposure are to greater or lesser degrees within people's personal controls, or at least they take personal responsibility for them, while they are involuntarily exposed to many industrial chemicals. As we shall show in our later chapter on 'Managing', people do not readily tolerate involuntarily-assumed risks, even if these risks are small.

Of course, the U.S. Congress and legislatures all over the world have passed many laws requiring the regulatory control of exposures to industrial products, whether they be present as environmental pollutants, in the workplace, or as ingredients in foods, medicines, and consumer products. Some of these laws single out carcinogens for special treatment, and thereby create heightened attention from regulators and the public.

So, the Doll–Peto estimates notwithstanding, it is necessary to continue to explore the scientific basis for being concerned about environmental chemicals that are carcinogenic. But as we move ahead on this topic we must keep in mind that we are dealing with only a piece of the total cancer problem, and are giving only cursory treatment to some perhaps overridingly important issues, such as the role of the diet and the issue of 'multiple factors.'

Identifying carcinogens – the epidemiological method

Epidemiology is the study of how disease rates vary among different populations and within the same population over time. Epidemiologists also attempt to identify the etiologies of those diseases. The Doll–Peto estimates are primarily based on epidemiological data.

Epidemiology has made the major contribution to our current understanding of the roles of individual environmental substances in human cancer causation, so that is why this space has been reserved for a discussion of the epidemiologic method and its limitations.

The obvious advantage of the epidemiologic approach over the animal test is the fact that it is based on observations in *Homo sapiens*, the species of interest. It is for this reason that results from epidemiologic investigations, when they are convincing, are given greater weight than results from animal tests. The decision on whether and to what degree results from epidemiological studies are given greater consideration than those from animal tests turns on several factors – simply put, upon an evaluation of just how convincing the results are at demonstrating the presence or absence of a conclusion about the existence of a cause–effect relationship. This is a far from trivial matter and decision-making is tricky, whether epidemiology studies reveal the presence or the absence of a relationship between a certain exposure and excess rates of human cancers. In the case of a 'negative' outcome it is necessary to judge whether an effect might not have been missed because of some study limitation. And in the case of a 'positive' outcome a judgement needs to be made about whether a cause–effect relation has not been falsely arrived at.

The obvious disadvantage of epidemiology studies, at least for identifying carcinogens, is that they can not be performed until after human exposure has occurred, and to be meaningful, until after it has occurred for several decades (some cancers do not develop for 30–40 years from the time of first exposure). Moreover, a study can not be performed until a population of individuals with certain characteristics – to be described below – can be identified and matched in certain ways to what is called a control population. Identifying such groups is not an easy task.

Epidemiology studies are generally capable of detecting fairly sizable risks but can usually not rule out small risks that may nevertheless be of public health or regulatory significance. Some experts suggest that unless a risk is large enough to be detected epidemiologically, then it is, almost by definition, not a significant public health matter. We shall point out in the chapter on managing risks that regulatory agencies have sought to reduce risks from environmental chemicals to levels far below those that can be detected using the epidemiologic method. Regulators have based these policies on their interpretations of the requirements of laws passed by Congress, which generally seek a high degree of public health protection.

With these general considerations in mind let us move to an explanation of the methods of epidemiology.

Descriptive epidemiology

Some useful information can be obtained by what is termed the descriptive method. Here the epidemiologist is searching for patterns of cancer in populations and seeking to determine whether those patterns can be correlated with environmental exposures or other factors. By correlation it is meant that a statistically discernible relationship can be identified between two factors. There is, for example, a reasonably high correlation between the level of aflatoxin intake in certain African and Asian populations and the rates of occurrence of liver cancer. This was determined by collecting data on the aflatoxin content of diets in certain areas, and assembling whatever data were available on liver cancer rates. The two factors can be plotted on a graph (liver cancer rate vs. daily aflatoxin intake), and statisticians can determine from the graph how strongly the two factors correlate. In this case the correlation is a relatively strong one. This is typical of a descriptive epidemiology study. Other types might include preparation of detailed maps showing regional variations in the rates of occurrence of specific types of cancers, or rates of death from these cancers. Sometimes such mapping will reveal 'hot spots' – regions in which there are unusually high rates of occurrence of certain cancers. For example, high rates of lung cancer were identified in certain coastal areas of Georgia. Close investigation revealed that these were also major ship building areas during World War II. Closer inspection suggested a correlation between asbestos exposure and this lung cancer 'cluster', because asbestos material was extensively used in shipbuilding.

These types of descriptive study are useful but are rarely definitive. Correlations by no means provide confirmation of causation. The Georgia lung cancer cluster is as strongly correlated with ship production as it is with asbestos usage; one would not conclude from this that shipbuilding 'causes' lung cancer. We know that asbestos causes lung cancer primarily because of results from other, *analytical* epidemiology studies, to be discussed below, and not because of the so-called correlational study. The rates of coronary disease in various countries correlate with annual television production figures. Again, it would be absurd to conclude from this that TVs cause heart disease. It is more likely that TV production correlates with some other set of factors that are causally related to heart disease rates.

Case-reports, which were mentioned in Chapter 4, constitute another type of human 'study,' although it would be incorrect to classify them as epidemiological investigations. We discussed in an earlier chapter their utility in identifying acute forms of toxic response. They have also yielded some useful data concerning carcinogens. Reports from observant clinicians helped identify the causes of certain forms of liver cancer in vineyard workers (from spraying arsenic solutions), and certain chemical workers cleaning vats containing residues of vinyl chloride. Unusual numbers of leukemia cases were reported among artisan Turkish shoemakers using large amounts of benzene as a solvent. These various case reports led to more thorough analytical studies, and all three of these suggestions have been confirmed by such studies. So the case-report has some utility, especially if the form of cancer is unusually rare (as in the liver cancers seen in vinyl chloride workers) or the exposures unusually intense (as in the case of the Turkish shoemakers).

There are several other types of descriptive epidemiology studies, but we need not say much more about them here. They will continue to be used to uncover major disease trends and to suggest hypotheses about cause–effect relationships. Full study of such hypotheses must be undertaken using more analytical techniques.

Analytical epidemiology studies

Analytical studies, of which there are two main types – case–control and cohort – are those in which the epidemiologist attempts to set up something approximating a controlled experiment.

Case–control studies are those in which individuals having a specific type of cancer are identified. If they are alive, these individuals can be asked questions concerning various exposures they may have experienced, in their places of work or in their daily lives. The same questions are asked of selected individuals of the same ages and sex, but who do not have the cancer. If the epidemiologist is dealing with individuals who have died of cancer, then other means are used to collect exposure histories. Results are compared to determine whether the 'cases' have experienced particular exposures or exposure levels not experienced by 'controls'. In concept, and sometimes in execution, case–control studies are not particularly complex. A particular advantage of case–control studies is that it is possible to take into account the influence of factors, such as smoking habits, that are known to influence cancer risks. A number of chemical carcinogens have been identified through the case–control method; most have involved workplace exposures or uses of certain medicines.

The case–control method does have its problems. Characterization of past exposure experiences by both cases and controls is sometimes based on memory. Individuals will be able to recall their occupations, but may not be able to pinpoint specific chemicals. They almost certainly can not specify the quantity of exposure (chemical concentrations), although they may perhaps recall the total time they were exposed. Study interpretation is sometimes complicated by the possibility that cases and controls may differ in their recollections of past exposures; if this is the case (and it's not easy to tell if it is), the study may not meet the criteria expected for a truly controlled situation. In some cases recollection can be checked by turning to documentary records of those exposures, but past record keeping on most occupational exposures was typically irregular and incomplete. Other problems exist in selecting appropriate controls, and there are sometimes serious technical difficulties associated with statistical analysis of study results.

Epidemiologists analyze results from case–control studies using something called the 'odds-ratio' (OR). This ratio is best defined by an example. Suppose our epidemiologist is inquiring into the possible causes of leukemia deaths among former employees of a rubber manufacturing facility; the solvent benzene is suspected, because of the case reports from Turkey regarding excess leukemias among shoe workers heavily exposed to this solvent. Company personnel records reveal a total of 17 leukemia deaths among former employees. A comparison group, consisting of two current or former company employees who do not have leukemia or did not die from it, for each worker who did, is selected by the epidemiologist to act as the 'control.' Work histories of the leukemia 'cases' and the non-leukemia 'controls' are reviewed, and a table (called a 2 × 2 table) is assembled:

Exposure to Benzene	Cases: Death due to leukemia	Controls: Not known to have leukemia
Yes	7 (a)	7 (b)
No	10 (c)	27 (d)
Total	17	34

The OR is computed from this table as follows:

$$OR = \frac{a \times d}{b \times c} = \frac{7 \times 27}{7 \times 10} = 2.7$$

An OR computed by this means would be statistically indistinguishable from 1.0 if there were no association between benzene exposure and leukemia deaths, but if it were statistically greater than 1.0, an association would be suggested. We emphasize 'suggested,' and avoid the word 'proved.' Evidence provided by an increased OR from a single study is generally inadequate to establish causal links, because of the limitations in case–control studies already mentioned. How such evidence contributes to an understanding of causal links will be discussed below, following review of the other major type of analytical study, the so-called *cohort* study.

Here the epidemiologist looks at a group of individuals – the cohort – known to be exposed to a suspect agent, and compares its health status, or the causes of death within it, with a control group. The control group is selected to be as similar to the cohort as possible, with respect to factors such as age, sex, race, geographical location, smoking habits, etc., except for the fact that its members have no exposures to the suspect agent (or at least much less intense exposures). This sounds a bit like a controlled study, but it is generally only approximately that. Remember the epidemiologist is attempting to take advantage of an existing situation; he or she is not setting up a laboratory experiment. Exposed and non-exposed groups are 'selected' to be as closely matched as possible, but the type of very close matching done in a laboratory experiment can not be achieved.

In any case the epidemiologist tries to match, and then worries a lot about uncontrolled factors that may confound interpretation of the study results. The numbers of new cases of disease or deaths per unit of exposed and non-exposed population, per unit of time – the disease incidence or mortality rates – are obtained and compared for the two populations. This comparison yields the Relative Risk (RR). Elevated RRs, as with elevated ORs for the case–control studies, suggest a relationship between exposure and disease.

Identifying a control group for which the appropriate incidence and mortality data are not available is, unfortunately, not at all uncommon. Instead of giving up entirely, the epidemiologist turns to data that may be available for the general population (for example, for the State or County in which the exposed cohort is located). In such a population the number of individuals exposed is typically much smaller than the number not exposed, so that disease and mortality rates observed in the general population may be expected to approximate those in a selected, unexposed (control) group. When general population data are used as the basis for comparisons, the epidemiologist calculates what is termed the SMR – Standardized Mortality (or Morbidity) Ratio. Numbers of deaths (or cases of disease) in the exposed group are divided by the numbers *expected* in that group if they had been at the same risk as the general population; by convention, the result is multiplied by 100, so that SMRs above 100, like RRs and ORs greater than 1.0, suggest the experienced exposures are involved in the observed mortality or morbidity rates.

Cohort studies, which can be either retrospective or prospective in nature, have at least one significant advantage over case–control studies. In case–control studies we are investigating a single disease. In the cohort study the epidemiologist has a chance to learn all of the diseases an exposure may lead to. Bias is somewhat reduced in cohort studies because classification of individuals into 'exposed' and 'unexposed' groups cannot be influenced by any prior knowledge of the disease states of the individuals (this knowledge can affect assignments of exposure categories in case–control studies).

Cohort studies are considerably more expensive than case–control studies, and may require long periods of time before enough cases of disease or death show up in the cohorts to allow meaningful statistical analysis.

Confounders

Cigarette smokers face an increased risk of lung cancer, and a few other types of cancers as well. In studies in several occupational settings, workers exposed to high levels of arsenic experience increased risks of lung cancer. Some workers also smoke cigarettes. How can the epidemiologist be sure that the excess of lung cancer cases among these several groups of arsenic workers is related to their exposure to the metal and not to their smoking habits? How can the potential *confounding* effects of smoking be ruled out? If the epidemiologist can be reasonably certain that the pattern of smoking among members of the occupational cohort approximately matches that of the selected control populations, then it would seem appropriate to conclude that smoking is not a significant factor in the interpretation of the role of arsenic exposure in the production of the excesses of lung cancers. If, however, there are reasons to believe that the pattern of smoking in the two groups is not similar, then it may not be possible to rule out at least a partial role for smoking. If it is not possible to adjust for differences in smoking patterns, the association is said to be confounded, and the epidemiologist can not be certain whether arsenic exposure, smoking, or both, contributed to the excess disease. Potential confounding factors need to be substantially ruled out before the possibility of causality can be taken seriously. The problem is particularly severe for diseases that are common and that are likely to be related to several different factors. Isolating the role of one among several possible factors when all are present is a formidable obstacle for epidemiologists, and various methods, some relatively complex, have been developed to attack the problem.

Causality – weighing the epidemiology evidence

Because individual studies are hardly ever controlled in all respects, and because interpretations of them may be complicated by a number of factors, epidemiologists rarely, if ever, reach firm conclusions based on results from single studies. (Unfortunately, this is not always true for members of the media.) Even where a particular type of cancer appears to be associated with a particular exposure situation, it is generally not appropriate to claim that a true causal relationship has been established. Before concluding that exposure to a particular chemical or group of chemicals is a causative factor, epidemiologists seek several types of confirmatory data.

First, they will usually await results from more than one study. In the ideal situation, results will be available from studies involving different groups of individuals who have been studied using different methodologies. If the same types of associations exist in all or many of these studies, the existence of a causal relationship gains in plausibility.

A second factor to be considered is the strengths of the associations observed – the magnitudes of the RRs, SMRs and ORs, and the degree to which statisticians can be sure that the observed values differ from the values that would have been obtained had the association been one of pure chance.

Does the risk, by whatever measure, increase with increasing exposure? If it does not do so, a causal relation can not be ruled out, but it surely becomes doubtful. Does risk increase with increasing time since first exposure – with what is termed latency? It would be odd and suggestive of a lack of causal connections to find higher cancer risks in workers a few years after a chemical exposure began than in workers whose exposure began several decades earlier (all other things being equal).

Biological plausibility helps. Does a role for the suspect agent in the observed disease make sense in the context of what biologists now understand about how diseases are brought about? Epidemiologists have to be careful with this one, because biological science is hardly capable of explaining the etiology of most diseases; so epidemiologists should perhaps not be overly hasty to rule out a possible cause because no known biological mechanism can be summoned up to explain the relationship. On the other hand, other biological data – for example, data from animal toxicity or carcinogenicity studies revealing that the suspect agent can cause the same effect in laboratory settings – can serve to strengthen epidemiological observations.

Lack of reliable information on the nature, magnitude, and duration of exposure to the agents under study also plagues most epidemiological investigations, generally in a big way. Epidemiologists are best off when dealing with prescription drugs, where rates of administration are pretty carefully controlled and documented, but even here exposure information is not nearly as good as we would like it to be. When we look at the epidemiology studies relating to occupational exposures to inhaled arsenic, asbestos, and chromium, all in IARC's 'sufficient evidence' category, we find some documentation of worker exposure, but generally it rests upon data of uneven quality and completeness (we are referring here mostly to historical exposures; current OSHA regulations and good industrial hygiene practices require frequent and careful checks on worker exposure levels). Although epidemiologists can often 'reconstruct' historical worker exposures sufficiently well to develop a plausible range of exposures, it is generally not a terrifically certain one. So, although in making an evaluation of causality, epidemiologists would like to know the relationship between worker risk and exposure – the dose–response relationship – they often have to settle for only a crude approximation of it, and in some cases they know nothing at all. This will be the case until epidemiologists develop and implement means to obtain different and more accurate measures of exposure; some ingenious techniques for doing so are now being explored.

Judging whether a causal relationship exists between a given exposure situation and a disease is, it should be obvious, far from straightforward. It is by no means a simple statistical exercise – the statistician can reveal only how likely or unlikely it is that an observed association is due to chance – and requires a high degree of expert judgement. The criteria used by experts – consistency of the observed associations, biological plausibility, dose–response properties, experimental reproducibility, absence of confounders, strengths of the statistical associations – are widely agreed upon, but can not be applied as a set of simple formulas. Scientific consensus or near-consensus, such as that sought by the IARC working groups, is generally necessary before there is broad acceptance of conclusions based on epidemiological investigations. And even if the 'mainstream' community of epidemiologists is of one mind on a particular agent, there will almost always be found some experts who disagree with those mainstream conclusions. This is not unexpected given we are dealing with an observational, not an experimental science.

A final item concerning epidemiology studies needs to be surfaced, if it is not already apparent. Inspection of the epidemiology data supporting the IARC classification of chemicals and chemical processes as carcinogenic shows they derive from studies of occupational exposures or medicines. Few environmental exposure situations have yielded results that contribute significantly to our classification of chemicals as known or probable human carcinogens. The obvious explanation for this is that occupational settings typically involve much more intense exposures (certainly this was the case in the past, and it remains true, but to a lesser degree, in the present), and so risks, if they exist, are likely to be relatively high compared with those expected in the broader, general population. The higher the risk, the easier it is to detect. Intake of medicines can also be relatively high. Exposures for these two settings can be more easily tracked than it can be for the general population. The only substances in Table 4 for which the epidemiology data did not derive primarily from studies in workers and patients are tobacco smoke and smokeless tobacco.

One interpretation of these observations is that people exposed to typical, low environmental levels of these same substances are not at risk. This is indeed possible, but it is important also to recognize that it is at the present time very difficult and perhaps impossible for epidemiologists to acquire meaningful data about low risks in large populations. We need always to distinguish whether we are talking about the true absence of an adverse effect, or the absence of studies sufficiently powerful to detect such an effect if it in fact existed. So, for every epidemiological outcome that is negative – that does not reveal an association between exposure to a chemical and excess risk – it is important to ask: how great a risk could have been missed, given the statistical power of the study? Differences in study outcomes are often explainable by differences in study methods and detection power; they may not reflect real inconsistencies. This is not to say that real inconsistencies do not exist, but only to note that it is incorrect to conclude that results are inconsistent simply by noting that some study outcomes are 'positive' and some are 'negative.'

The problem of apparent as against true inconsistencies also comes up when comparisons are made between the results of epidemiology studies and experimental animal studies. 'Dinitrochickenwire (DNC) causes cancer in rats, but there is no evidence that it does so in humans.' This statement could be uttered for several hundred chemicals. But before it is accepted at face value, it is critical to inquire whether the speaker means that no epidemiology data exist for DNC (in which case the speaker has committed a sin of omission), or whether he means that epidemiology studies have been conducted and are negative in regard to causation. If the latter is the case, it is important to inquire further whether the studies were sufficiently powerful to rule out a risk the size of the one that DNC produced in animals. If they were sufficiently powerful to do so, then it might be concluded that DNC is not a human carcinogen. But, if they were not sufficiently powerful – and, unfortunately, this is often the case – then we should be reluctant to toss out the positive animal data. As we emphasized in Chapter 6, animal toxicology data, obtained from appropriately designed and conducted experiments, while not adequate to establish that an agent can cause disease in humans, appear to be sufficiently telling that to reject them without good scientific reasons would be imprudent. Because, in fact, most of our knowledge about chemical carcinogens derives from animal studies, we devote the next several sections to them.

Animal bioassays for carcinogens

Most operational definitions of carcinogens include reference to animals as well as humans. As two eminent experts, Gary M. Williams and John H. Weisburger of the American Health Foundation, concisely put it:

Chemical carcinogens are defined operationally by their ability to produce neoplasms. Four types of response have generally been accepted as evidence of induction of neoplasms: most importantly (1) the presence of types of tumors not seen in controls; (2) an increase in the incidence of tumor types occurring in controls; (3) the development of tumors earlier than in the control; and (4) an increased multiplicity of tumors. For agents producing any of these effects, the term 'carcinogen' is generally used, although it literally means giving rise to carcinomas, i.e., epithelial malignancies. For agents producing sarcomas of mesenchymal origin, the term 'oncogen' is more correct.[10] As evidence of carcinogenicity, production of even benign neoplasms is accepted, a practice justified by the fact that no chemical has yet been identified that produces exclusively benign neoplasms.

Although some experts do not accept that benign tumors should be counted as evidence of carcinogenicity, regulatory agencies and bodies such as IARC generally do so, for the reason suggested by Williams and Weisburger, and also because they hold that cancer development is a multistep process and that benign neoplasms represent one stage of the process, the next being the progression of those tumors to malignancies.[11]

[10] Epithelial cells are those that line tissues and organs; mesenchymal refers to connective tissues, blood vessels, and lymphatics.
[11] Whether a tumor is benign or malignant is obviously of immense importance to a physician attempting to find the proper treatment for a patient (to say nothing of its importance to the patient). Identifying chemical carcinogens is a different matter.

Animal studies, or cancer bioassays, are used to identify carcinogens. As we have said, several hundred chemicals have been so identified, and many are under test at this very moment. Some chemicals are selected for testing because certain laws require that manufacturers do so. In many cases there may be no clear legal mandate that manufacturers conduct such tests, so that the task has been taken on by the federal government, largely by NTP, as noted earlier.

Chronic exposure is needed to identify most carcinogens. The cancer bioassay is thus similar to the chronic toxicity test covered in the last chapter. The major difference is that the toxicologist and the pathologist are generally searching only for evidence of tumor formation, and certain of the clinical measurements typically associated with the chronic toxicity study may be omitted from the protocol.

Under currently used protocols, animals are assigned to one of four groups. One group receives the MTD, the estimated *maximum tolerated dose*, a second receives one-half of the MTD, and a third receives one-fourth. Controls make up the fourth group. (Just what is meant by the MTD will be reviewed later.) Approximately 60 males and 60 females are assigned to each group. Both rats and mice are used. Study duration is the rat and mouse lifetime, about two years. Variations on this design are possible, although some factors, such as the use of MTD and fractions thereof, and lifetime dosing, are generally considered essential. Total time from study planning to final report is typically 3 or more years. All other necessary study features mentioned in Chapter 6 in connection with animal bioassays are built into the cancer bioassay.

At the end of the study animals are killed, necropsied, and taken to the pathologist's laboratory. Animals dying during the study are also taken for examination, assuming their deaths were noted sufficiently soon to avoid decomposition of their tissues.

All organs and tissues are examined for clearly visible lesions – the so-called gross examination. Tissues are preserved, embedded in paraffin, sectioned, placed on glass slides, stained in various ways to assist identification of cellular components, and put under the microscope. For each animal more than 30 tissues and organs are prepared for examination. Given the numbers of animals, dose groups, and tissue sections per animal, a total of about 40 000 individual slide readings are made by the pathologist for each chemical subjected to bioassay.

The pathologist's diagnoses are tallied and assigned to the appropriate animals and groups. The statisticians are then put to work to determine which, if any, specific neoplasms occur with significantly greater frequency in treated groups compared with control groups. The total data base, together with the outcome of the statistician's work, is then reviewed, typically by a group of toxicologists, pathologists and statisticians.

Assuming there are no problems with study design or conduct, which is not always the case, the expert group is looking to answer the following questions:

At which sites (tissues and organs) do excess tumors occur in treated (dosed) animals?

Is their rate of occurrence significantly different from their rate of occurrence in control animals, in the statistical sense?

Does their rate of occurrence increase with increasing dose?

What is the biological significance of the observed excesses, irrespective of their statistical significance?

Are the excesses mostly benign or malignant tumors?

Do the excesses occur in only one body site of one sex of one species, or (at the other extreme) do they occur at multiple sites in both sexes of both species?

Do some tumor types occur much earlier in treated than in control animals?

Depending upon the answers to those questions, the experts will judge the evidence of carcinogenicity as negative or as more or less positive. A negative classification will be assigned if there are no tumor excesses of biological or statistical significance. Positive outcomes will be judged as very convincing if tumors occur at many body sites in both species and sexes, are mostly malignant, show strong responses, with risks increasing with increasing doses, and with relatively rapid onset. Weakly positive responses might be those restricted to a single body site in one sex of one species, with only moderately or weakly elevated rates of tumor formation, and with relatively few malignancies showing up. Most outcomes are between these extremes.

Table 6 shows the outcome of a carcinogenicity bioassay for a chemical called ethylene thiourea, a decomposition product of a certain class of fungicides that can be found in treated foods.

Several features of the carcinogenicity results on ethylene thiourea deserve discussion. Not all of the animals in a given dose group developed thyroid tumors over the course of their exposures; only a fraction of each group did. Another name for this fraction is *lifetime tumor incidence*; still another is *lifetime risk (probability)* of tumor development. Second, the lifetime risk of tumor development, or, since in the definition of carcinogenesis 'tumors' were equated with 'can-

Table 6. *Cancer bioassay results for ethylene thiourea (ETU)*
(Graham *et al.* 1975. *Food and Cosmetics Toxicology*
Vol. 13, p. 493)

Dietary level of ETU (ppm)	Fraction of rats developing thyroid tumors over their lifetimes
0 (Control)	2/72
5	2/75
25	1/73
125	2/73
250	16/69*
500	62/70*

* Statistically distinguishable from controls.

cers,' the lifetime risk of cancer increases with increasing dose. This dose–response relationship, where 'response' is extra lifetime cancer risk, is typical of carcinogens. The matter of dose–response relations is treated extensively in Chapter 9.

Also interesting is the fact that the incidence of thyroid tumors was statistically distinguishable from the incidence in controls (which was not zero!) only at the 250 and 500 ppm dietary levels. This might suggest there is no extra cancer risk from this chemical at dietary levels up to 125 ppm and perhaps at even higher levels, up to somewhere below 250 ppm. Before this conclusion (which may be valid) is accepted, the discussion of Chapter 6 regarding the power of these types of animal tests to detect risks below about 1 in 10 needs to be recalled. Statistically, the best that can be said about the result at 125 ppm is that the cancer risk does not exceed some low level, which can be calculated by a statistician; it is incorrect to conclude that it is known to be zero. Other biological data may, however, suggest that 'zero' or something very close to it is a reasonable conclusion; but the bioassay result alone is insufficient to permit this conclusion.

Other tests

Epidemiological and lifetime animal bioassays are the two major means for identifying chemical carcinogens, and, as has been observed and explained, most carcinogens have been identified through the animal test. Other means exist for identifying carcinogens, but none has yet achieved the same standing among scientists. There are, for example, a number of so-called 'short-term tests,' some involving animals and others utilizing cells from animals or microorganisms, that provide, with greater or lesser degrees of certainty, a signal about possible carcinogenicity. A set of these 'short term tests' includes procedures for detecting the capacity of a chemical or its metabolites to damage or alter the functioning of cellular genes. As shall be seen in the next chapter on 'Mechanisms,' gene damage is thought to be an important event in the genesis of cancer, so it will be necessary to go into *genotoxicity* studies in a little detail when embarking on that discussion. But with respect to the issue of 'short term tests' for identifying potential carcinogens, and the role of tests for genotoxicity, no more need be said here, because none is now used for greater purpose – certainly not for human risk assessment – than pinpointing chemicals that should be sent through the more definitive, long-term bioassay.

Chemical structure

What has emerged from nearly a century's study of chemical carcinogens is that certain molecular structures appear to be riskier than others. Many specific members of certain classes of organic compounds, grouped together because of structural similarities, have been shown capable of inducing excess neoplasms, thus raising suspicions regarding the entire chemical class. PAHs and aromatic amines have already been seen as suspect chemical types. Representatives of a few more suspect classes are shown on the next page.

Although any chemical containing an azo group (-N=N-) or nitrosamine group (-N-N=0) is suspect, not all members of the class turn out to be carcinogenic when tested; this is true for all suspect classes. Sometimes other structural features in the molecule serve to mitigate the effect of the dangerous group, for example, by helping the body rapidly to eliminate the compound and its metabolites.

Many other types of organic compounds have been shown to induce excess cancers. The structures of some of the more interesting of these are depicted on page 136, along with a notation about their origins.

Toxicologists have begun to understand how the chemical structures and some of the other properties of these agents contribute to their

Nitrosamines

Azo compounds

$$CH_3-N-CH_3$$
$$|$$
$$N=O$$

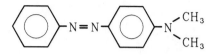

Dimethylnitrosamine

4-Dimethylaminoazobenzene
('Butter yellow')

Halogenated hydrocarbons

Carbamates

$$\begin{array}{cc} H & H \\ \diagdown & \diagup \\ C=C & \\ \diagup & \diagdown \\ H & Cl \end{array}$$

$$CH_3-CH_2-O-\overset{\overset{\displaystyle O}{\|}}{C}-NH_2$$

Vinyl chloride

Ethyl carbamate

carcinogenicity. This topic is central to the subject of Mechanisms, the subject of the next chapter. But there are some additional issues needing review before that interesting matter is explored.

Are animal bioassay results to be taken seriously?

Yes, but with caution. Reference has been made several times to the fundamental biological similarities of mammalian species, and to the expected similarities in response to chemical toxicity in animals and human beings. These expectations have been borne out in a large proportion of those cases in which there has been an opportunity to obtain toxicity data in both humans and animals, so that it would be imprudent to ignore the results of cancer bioassays. At the same time these results need to be carefully scrutinized, because they can easily mislead.

All known human carcinogens – the substances ranked by IARC as having been causally linked to human cancers – have been shown to be capable of inducing cancers in some (but not all) species of experimental animals, with the possible exception of arsenic. Arsenic is a human carcinogen, however it has not been adequately tested in animals – so it is perhaps not a real exception to the rule. A few examples of

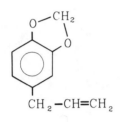

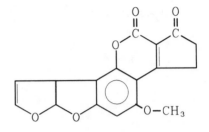

Safrole
(natural plant product,
once used as flavoring
agent)

Aflatoxin B₁
(mold product, see Prologue)

Cl–CH₂–O–CH₂–Cl

Bis (chloromethyl) ether
(Industrial chemical)

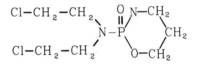

Cyclophosphamide
(anti-cancer-drug)

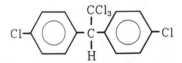

DDT
(Insecticide, cancer promotor,
see Chapter 8)

O
‖
CH₃ –N=N–CH₂ –OH

Cycasin
(Appears, bound to a
sugar molecule, in the
cycad nut)

carcinogens that are known to be active in both humans and animals, are presented in Table 7.

It is of more than a little interest to note that the sites of tumor formation do not always match across species. Benzidine, a substance once widely used in dye manufacture, was shown many years ago to be a carcinogenic risk for the bladder in workers exposed to excessive levels. The rat bladder is not responsive to this substance, but its liver is. It wasn't until Wilhelm Hueper turned to the dog that bladder cancer could be reproduced in a laboratory animal. It is now understood that benzidine metabolism is similar in dogs and people, and that

Table 7. *Some chemicals known to be carcinogenic in humans and their sites of action in animals*

Chemical	Carcinogenic sites in humans	Carcinogenic sites in animals
Aflatoxins	liver (suspected only)	Liver in rats, mice, monkeys, and several more species
4-Aminobiphenyl	bladder	Bladder in mice, rats, rabbits, dogs
Asbestos (inhaled)	lung, mesothelium	Lung and mesothelium in mice, rats, hamsters
Benzidine	bladder	Liver in rats. Bladder in dogs
DES	vagina	Cervix, vagina, other tissues in mice, rats, hamsters, monkeys
2-Naphthylamine	bladder	Bladder in hamsters, dogs, monkeys Liver in mice, rats
Vinyl chloride	liver	Lung, mammary gland in mice Liver, kidney in rats

metabolism in the rat takes a different course. It is also understood that certain benzidine metabolites, and not benzidine itself, are the proximate causes of tumors. Knowledge of metabolic differences helps explain the species similarities and differences in tumor response.

Empirical information of the type presented above seems to fit theories about the biological similarities of various animal species, including our own, and where differences occur, as with benzidine, it seems that explanations consistent with current understanding of biology are available. So we ask, should we accept as incontrovertible that every animal carcinogen is a potential human carcinogen, when we have inadequate direct information regarding effects in humans? Several hundred animal carcinogens are known to us, and a significant fraction of these can be found in the environment and a larger fraction in the workplace. Should all of these be considered cancer threats to humans?

Let us skip by the question of the adequacy of the animal tests used to identify these agents. The general quality of the animal test is obviously of great importance in the overall evaluation, as was stressed at several points in Chapter 6, and these questions can not be ignored in the case of cancer bioassays any more than they can in any other type of toxicity test. But the more interesting questions arise when we move beyond the question of study quality.

If we know nothing else than the facts presented thus far – that mammalian species exhibit the same basic biological characteristics (although some differences exist), and that all known human carcinogens are also known to be active in at least one animal species – it would seem foolish to ignore or downplay positive animal tests for other carcinogens, even when no telling human data are available. Indeed, regulatory policy in the United States and the rest of the world embraces animal test data for inferring hazards to humans. But how much of this is science and how much is simply a matter of prudence in the absence of scientific certainty? Surely both are included.

As a simple matter of logic, the fact that all known human carcinogens have been found to be carcinogenic in a least one other animal species is, of course, not proof that every substance found carcinogenic in one or more animal species will be carcinogenic in humans. Logic also informs that even substances found incapable of producing excess tumors in adequate animal tests can not be absolutely rejected from the class of human carcinogens. We are, for both positive and negative outcomes, dealing with probabilities, not certainty.

What sorts of evidence might increase the probability that an agent is or is not a human carcinogen? Although it can not be proved empirically, it would certainly seem plausible, for example, that a substance producing large excesses of tumors at several sites in several species and strains of test animals and in both sexes, and at multiple doses, is more likely to be carcinogenic in human beings than one that produces only a small excess of tumors at a single site in one species and sex, and that produces no other excesses in other species and strains. Similarly, the greater the number of clearly negative outcomes in animal bioassays, the more convinced we become that the agent is not carcinogenic to humans. This type of weighing of the evidence is one step in the determination of the probability that a chemical is carcinogenic to humans.

Metabolism data might help. Evidence that a chemical's metabolic patterns in test animals are uniform among several species increases the chances that human metabolic handling of the chemical, if we could obtain information on it, will turn out to match that of animals, whereas the existence of substantial differences among animal species creates uncertainty about which species, if any, humans might match. This business gets complicated quickly, however, because for some chemicals substantial differences in at least rates of metabolism exist among members of the human population.

A particularly controversial aspect of judging the applicability to humans of animal test results concerns the fact that certain sites of tumor formation in certain species, strains, and sexes of test animals are suspected of being uniquely susceptible to carcinogens, or very nearly so, such that excesses at those sites observed in animal tests are considered of dubious relevance to human beings, at least by most experts. Perhaps the clearest example is the kidney of male rats. As the male rat ages his kidney naturally undergoes a predictable series of degenerative changes that seem not to occur in other species, including humans. The female rat kidney also undergoes degenerative changes as it ages, but the changes occur more slowly and are less severe than in males. Certain chemicals such as gasoline (actually a mixture of many hydrocarbons) are capable of accelerating the rates of those degenerative changes in male rats and of also increasing the development of a certain type of tumor in the male rat kidney. These same chemicals produce no such changes in female rats or in either sex of mice. Some biochemists and pathologists believe they understand the underlying biological reasons for these changes leading to tumorigenesis. And they appear to be unique to elderly male rats.

Some other tumor sites are similarly susceptible. The male mouse lungs and liver, for example, tend to develop high and highly variable rates of tumors, even when the animals are untreated with any agent. The reasons for this phenomenon are not entirely clear, although it appears to be due to unusually high populations of certain cells that have undergone what is termed 'initiation' into the carcinogenic process. Initiated cells, which are a topic for the next chapter, are unusually susceptible to the effects of certain types of chemicals, and progress easily to tumors when assaulted with high doses of those chemicals. The human liver and lungs, as well as these same organs in female mice and in rats, do not seem to contain the same heavy concentrations of susceptible cells, and so may not react so readily to an invasion of chemicals that promote male mouse lung or liver cells to malignancies. We should also mention that the pathologist's diagnoses are not entirely objective – an element of subjectivity enters, especially in the case of rodent liver tumors – so that disagreements about whether a particular lesion is truly a neoplasm arise with surprising frequency.

These types of issues may seem arcane, but they are at the heart of

many of the scientific debates that take place whenever a new carcinogen is identified and announced to the world. Each bioassay result is
unique and needs to be scrutinized with these and many other considerations in mind. This is obviously a job for experts from several
disciplines, and decisions are best made through discussions among
them and a consensus judgment; dissenting experts can almost always
be found, and this is not surprising to anyone who has developed a
thorough understanding of the relevant scientific issues.

MTD

The use in animal cancer bioassays of the estimated maximum tolerated dose, MTD, is perhaps the one issue that provokes the most
heated scientific debates, as well as the greatest public skepticism about
the meaning of animal data. First, what is the MTD? Second, why is its
use problematic?

Recall the earlier discussion about the limited detection power of
animal tests involving only a few dozen animals in each test group. The
group size is chosen, we have said, largely for practical reasons. The
sample size limits the excess tumor incidence that can be detected, and
statistically distinguished from the incidence observed in control
animals, to something like 5–10% (1/20–1/10) and higher. Tumor
incidence rates, or tumor risks, below this rather high level are not
detectable unless the numbers of test subjects are greatly increased.

Many of those scientists involved during the 1950s–1970s in the
design of animal cancer tests came to the view that this statistical
'insensitivity' had to be dealt with by administering high doses. The
notion was that the test animals would be administered the highest
dose they could tolerate and yet survive close to their full lifetimes.
Survival is critically important, because neoplasms tend to develop
very slowly, and if too many animals die early the opportunity to detect
tumors may be lost. The dose should thus be high enough to overcome
the statistical limitations associated with using small groups of animals, but not so high that an excessive number of early animal deaths
result from toxicity other than carcinogenicity.

This seemed to make sense and, in fact was adopted as policy by
regulatory agencies and the National Cancer Institute and continues as
their policy to this day.

But how is the MTD determined? The doses have to be decided in advance of the testing. The best way would be to conduct a lifetime dosing study over a wide range of doses, and then determine the maximum dose that satisfied all the MTD criteria; results from higher doses that were excessively toxic would be discarded. The lifetime study would then be conducted with the proper MTD. This is extremely expensive and time consuming. Instead, the practice has become to limit the pre-chronic testing phase to a study of subchronic duration, typically 90 days, and to estimate the chronic MTD from the subchronic results. This involves some educated guessing. Not infrequently the guessers miss the mark, but this can be known only after the cancer bioassay is long underway or completed. They frequently miss in one direction – they overestimate the MTD, and survival is reduced – although there are certainly cases in which the MTD is underestimated. Neither result is desirable. At the same time no one wants to discard data from studies that cost many hundreds of thousands of dollars and that take 3–4 years from start to finish. This is troublesome.

When the MTD is underestimated, then toxicologists worry that a negative outcome is not very convincing. The relative statistical insensitivity of the test may not have been overcome and a significant risk of tumor formation might have been missed. If the chemical is important enough – human exposure is or could be substantial – the test may have to be repeated at higher doses to provide better guarantees.

If the MTD overshoots the target, other issues arise. If a negative outcome is observed, the excess numbers of early deaths might be blamed, because their occurrence would have reduced test sensitivity – it might have been the case that an insufficient number of animals survived long enough to develop tumors. A dilemma arises not unlike the one that occurs when the MTD is underestimated.

Sparks really fly when the MTD is exceeded and excess tumors are nevertheless observed. Why is this problematic?

Many toxicologists think that interpretation of such an outcome as evidence for carcinogenicity is erroneous. They contend that the excessive toxicity that somehow decreased animal survival, or that made them excessively ill, contributed to the production of the extra tumors, and that in the absence of that toxicity, neoplasms would not have developed. In other words the tumorigenic response was not directly a consequence of the chemical, but rather arose from cells so damaged by toxicity that they were put at extra high risk of progressing to the neoplastic state. If human exposure to the chemical were clearly never to reach levels that could cause the overt, initiating toxic damage (and this is almost always the case), then interpretation of test animal results as potentially applicable to humans would be absurd.[12]

These arguments are countered by the point that it is difficult to be sure that what we have called the 'initiating toxic damage' was actually responsible for the production of tumors. It might still be the case that the neoplasms would have developed even in the absence of that toxicity. So we should not, this argument goes, drop our concern until we are certain that cancers would not have occurred without prior toxicity. This is not easy; it's tough to rule one way or the other on this issue without additional data.

In these circumstances regulators are usually more fearful of reaching 'false negative' than they are of 'false positive' conclusions. That is, when an uncertainty such as this one exists, regulators will typically choose to chance being incorrect by calling the test positive rather than to take the chance of being incorrect by calling it negative. They err on the side of safety. The manufacturer whose product is being threatened will obviously object, but will likely not be successful unless additional data can be brought forth to convince the regulators that exceeding the MTD created a highly artificial circumstance and a false conclusion about carcinogenicity.

A few instances of this phenomenon – high dose toxicity leading to tumors – seem fairly well accepted, even by regulators. Dr. David Clayson, of Health and Welfare in Canada, has shown, for example, that chemical induction of bladder tumors in the rat is sometimes a consequence of the chemical's capacity to produce stones that deposit in bladder tissue. The presence of those solid bodies creates the conditions for the transformation of normal bladder cells to malignancies. If the dose of the stone-producing chemical is dropped below that necessary to create stones, no neoplasms form. The toxic damage – stone deposition – somehow puts the bladder cells at extra risk; the underlying biology of this stone–cell interaction and its relation to carcinogenicity is moderately well understood, and the Clayson hypothesis is fairly widely accepted. Clayson holds that 'the action of any non-genotoxic agent that induces bladder tumor in rodents in the presence of stone is therefore most probably irrelevant to human carcinogen risk assessment' ('non-genotoxic' agents are discussed in the next chapter).

---

[12] We shall later see why a carcinogenic response, as against what we have called a toxic response observed at similarly high doses, might still be of relevance to humans exposed at much lower doses.

But in most cases detailed experimental studies to support the hypothesis for a role of toxicity in production of neoplasia are not available, so regulators rule with caution.

Even if the MTD is not exceeded there can be reasons to worry about positive outcomes obtained at it. Metabolism is very often a major factor in the production of toxicity and carcinogenicity. A chemical's metabolism may, however, undergo substantial changes, both in terms of the amounts of metabolites produced and even in the chemical natures of those metabolites, as the size of the administered dose is changed. In most cases the MTD is estimated from observations of toxicity over a range of doses administered in 90-day studies. ADME studies are only infrequently performed to assist estimation of the MTD. What if, unknown to us, the nature, pattern, or rate of metabolite formation at the selected MTD is radically different from that occurring at much lower doses? Might this not suggest that observations of excess neoplasms at or near the MTD resulted from metabolites that either do not exist or that are formed at much different rates when doses are very low? Even if the high dose metabolic profile results in no unusual toxicity (except excess tumors), such that the survival of the animals is not threatened, might not the high dose results be inapplicable to humans, or even to the same animal species, at low doses where the metabolic profile is greatly different? Indeed they could. In fact many experts, such as Robert Squire, an eminent pathologist at Johns Hopkins University School of Medicine, and Ian Munro, Director of the Canadian Centre for Toxicology, have called for the routine incorporation of ADME studies in the determination of the MTD so that this potential problem can be circumvented.

Again, however, regulators become cautious when they are not sure, and they will become convinced that the altered metabolic patterns are significant only with a clear demonstration that the excess tumors would not have occurred in their absence. Such a demonstration can not be made without additional and usually highly technical studies of metabolism and its relation to dose and to tumorigenesis.

A much longer list of issues relating to interpretation of animal cancer studies could be made and commented upon but it should by now be abundantly clear that unambiguous results are not common, that conflicting scientific interpretations are expected, and that in the regulatory setting most uncertainties are resolved by erring on the side of safety – by a tendency to assume the more pessimistic of two conflicting interpretations. At the same time it should be recognized that there is more agreement among toxicologists on these matters than might be implied from the discussion of the areas of possible conflict. Consensus is no doubt too strong a word to characterize the present state of affairs, but toxicologists do take animal data very seriously; the disagreements usually arise over the highly important details of study interpretation, and not over basic principles.

The opportunity to reduce the uncertainties regarding animal data and the broader questions of the role of environmental chemicals in human cancer and other diseases is at hand. Toxicologists are more frequently turning to the study of the mechanisms underlying the biological actions of carcinogens and other toxicants, and we now turn to this topic, to clarify some matters covered in the present chapter and to prepare the way for our later discussions of dose–response and human risk.

8

Mechanisms

Fill up some animals with chemicals and see what health damage ensues. Follow a population of people exposed to some chemicals and see if they get sick more frequently or earlier in their lives than do unexposed people. Put crudely, these are the objectives of animal testing and epidemiology studies. Such studies are highly important, no doubt, but do not provide much scientific understanding. Scientists should be inquiring into the biological and chemical processes underlying these observations. What course does the chemical take through the body – its ADME profile? What specific chemical, the compound entering the body or one or more of its metabolites, brings about the damage? What interactions between the toxic and natural biological molecules occur in tissues and cells, and how do these interactions lead to toxic injury, disease or death? What factors influence the evolution of toxic damage over time? In short, what are the *mechanisms* of toxicity?

Why is this important? To learn about mechanisms requires a lot of sophisticated and expensive scientific work. To engage in it is a great deal of fun for toxicologists, and can result in a long list of publications in the very best scientific journals. Basic biological knowledge can be enhanced; there are many examples in which biologists have learned something fundamental about biochemical and physiological processes by looking closely at how those processes are altered or impaired by chemicals and drugs. Medicine has profited enormously from the work of pharmacologists who have sought to understand the most intricate molecular details of the interactions of drugs with biological targets, both to produce their beneficial effects as well as their harmful side effects. Indeed, much of the current work on mechanisms of toxicity was inspired by the pharmacologist's study of the mechanisms of drug action.

In our context – attempting to understand whether and to what extent chemicals may pose human health risks – the study of toxicity mechanisms helps in at least two major ways.

First, it can help clarify whether effects observed in animals are also expected to occur in people exposed to a sufficiently high dose. Once the underlying toxicological mechanisms are in focus, the risk assessor is in a much better position to evaluate whether people are likely to be equally susceptible to the agent. If, as in the case of gasoline-induced kidney cancers in male rats, discussed in the last chapter, toxicologists come to understand that the mechanisms entail biochemical processes that are unique to male rats, they have greatly clarified their understanding of the potential risks to human beings. If it is observed that the dangerous metabolite of carbon tetrachloride is formed in human as well as in mouse and rat livers, confidence in extrapolating results from those animal species to humans increases greatly. The animal model is no longer a 'black box', in which all sorts of important events are taking place, hidden from us.

The other problem that an understanding of mechanisms can help resolve concerns the relevance to low doses of observations of toxicity at high doses. It should be clear by now that most animal tests and epidemiology studies are pretty much limited to high-dose–high-risk situations. The molecular toxicologist, by contrast, has tools available to study mechanisms at relatively low doses. So, in theory at least, there is an opportunity to learn whether the toxic mechanisms at work at high doses, those associated with observable toxicity, are also underway at low doses, where the ultimate manifestations of toxicity are not readily observable. The degree to which the low dose mechanisms match high dose mechanisms helps predict the degree to which low dose toxicity should match that seen at high doses.

Having said all this we should hasten to add that toxic mechanisms are, in general, poorly understood. It is probably safe to say that the mechanism of action of no chemical is understood in every detail. Toxicologists know a great deal about a few chemicals, a little about many, and next to nothing about most. The practical effect of this situation is that, at the present time, information on mechanisms generally has but a minor influence on the human risk assessment process. It is one of the fundamental beliefs of toxicologists that understanding of mechanisms is at the essence of the science, so the community will no doubt continue aggressively to promote their study and the incorporation of their results into the human risk assessment process.

Regulatory agencies, one of the principal producers of risk assessments, have shown much interest in this topic but have been reluctant to incorporate mechanistic information because it is generally incomplete in some way. This is not a particularly well-conceived position, I would argue. Regulatory risk assessments rely heavily upon the 'black box' animal model, where uncertainties abound but which are explicitly recognized in only a very broad way. The study of mechanisms reduces major uncertainties, but usually reveals additional, smaller ones. The policy that rejects the use of data on mechanisms because they are incomplete in some way, or contain uncertainties, yet which still embraces the very large uncertainties of the 'black box' model, seems wrongheaded. It also tends to discourage research on mechanisms. From the regulators' viewpoint, there are potential dangers in moving too quickly to incorporate mechanistic data. As it turns out, most information on mechanisms developed so far tends to suggest that risks to humans are less than might be understood from simple reliance on animal test results and the usual risk assessment assumptions used by regulators. To incorporate this mechanistic information, when not all questions are answered, might lead us erroneously to understate human risk in too many cases. The tension felt by regulators who want both to follow good scientific practices (which would encourage use of the understanding that data on mechanisms permits us to have) and to protect public health in the absence of a high degree of scientific certainty, needs to be appreciated. At the same time scientists need to encourage the regulators to move toward greater reliance on the achievements of the research community. Of course, we all need to keep in mind that 'scientific certainty' is an illusory goal.

Let us inquire in the remaining sections of this chapter into some of the current insights provided by the study of mechanisms. Almost all of the discussion will concern carcinogens, but many of the principles apply to agents producing other toxic responses. Experimentalists in the field of chemical carcinogenesis have been at the study of mechanisms for five or six decades, and have gained a phenomenal understanding of what has turned out to be an extraordinarily complex business. We can only scratch its surface.

A model of carcinogenesis

As a matter of exposition it is probably easiest to begin with a description of what are currently considered to be the major events in the development of malignancies from normal cells. The picture presented captures the process in only a very broad way, and dozens of intricacies will be omitted. This is justified on two grounds. First, scientific consensus on the model might be said to exist only at the broad level; scientists are still struggling to understand and agree on details. Second, to create much understanding of details requires that we first delve pretty deeply into basic molecular biology, cellular genetics, and the fundamentals of cellular growth and tissue development; this is too much of a diversion from our central topic. We shall, however, get to review some of the biological underpinnings of the model and to a discussion of its implications for risk assessment.

It has seemed pretty clear for several decades, from both studies in humans and in experimental settings, that carcinogenesis is a *multistage* process. At the broadest level, the process can be thought of as one in which a normal cell is first converted to a permanently deranged cell, which is called a neoplastic cell, and a second sequence in which the neoplastic cell develops into a tumor, a neoplasm, that the pathologist can observe: neoplastic conversion and neoplastic development.

How does the neoplastic cell come about? It seems pretty clear that the *initiating event* is brought about when the chemical carcinogen, which is in many, if not most, cases a metabolite and not the administered chemical, reaches a cell's nucleus and chemically reacts with DNA, the genetic material. This reaction constitutes DNA damage, an unwelcome event because this magnificent molecule controls the life of the cell and the integrity of its reproduction. Fortunately, cells have a tremendous capacity to repair DNA damage; these repair mechanisms have been at work probably since life began to evolve, because most types of cells have always been assaulted by DNA-damaging radiation and chemicals from many natural sources. If some of the damage is not repaired, and this happens because repair is not 100% efficient, and the cell undergoes replication when the damage is present, then the damage is passed on to the new cells, and can become permanent – a *mutation* has occurred. These mutations can take several different forms and result in different types of cellular alterations. Of particular interest these days is the role of so-called *oncogenes*. Oncogenes are the deranged twins of their normal counterparts, the protooncogenes. The latter are pieces of DNA that direct the synthesis of proteins necessary to the normal operation of cells. Sometimes, for reasons that are poorly understood, the functioning of the protein becomes uncontrolled. When this happens, the protooncogenes are said to have mutated to oncogenes and the cell in which this occurs begins its journey to a neoplastic transformation.

Experts refer to all cells that have been so altered in their genetic features as *initiated* cells. The initiated cells might be fully neoplastic, in which case their proliferation takes place in a wild, uncontrolled fashion, or more commonly, they are only partially neoplastic. In the latter case the abnormal cells may still be under some control, and held in check by the actions of certain biological factors inherent to the organism in which the neoplastic conversion is occurring (the host). But some proliferation of these abnormal cells may also occur, brought about perhaps by continued chemical assault. Cell killing by chemically-induced toxicity may cause tissues to produce more cells at faster rates, i.e., to *proliferate*, and during this proliferation phase the abnormal cells can be further converted to fully neoplastic cells because of 'genetic errors' that also proliferate – rapidly proliferating cells are at increased risk of this type of error.

Neoplastic cells may remain unobtrusive, under the regulatory control of various host factors. But when these controls, which may involve some intricate molecular communications between cells, break down, the abnormal cells can begin to grow and develop. The process is enhanced by chemicals called *promoters*, about which more later. After this *progression* to full-fledged malignancies takes place, usually by way of what pathologists refer to as the 'benign' state. Promotion and progression are thus the two processes involved in creating a cancer out of a neoplastic cell. The ultimate neoplasm is thus a population of cells that arose from a single cell, what biologists refer to as a population of clones. The monoclonal origins of cancers is suggested by many studies of human and animal cancers.

This broad picture is considered pretty accurate by most cancer experts. Multiple stages are involved. They take place at different rates, and whether and at what rate they occur depends upon many factors, including at least:

(1) The concentration over time of the initiating carcinogenic chemical (typically a metabolite) at the cellular target.
(2) The presence of chemicals, which might be the carcinogen itself, its metabolites or even some other chemical – including some normal dietary

components – that may restrict or enhance conversion or development, at
several different points of the process.

(3) The influence of host factors, including cellular genetics and factors such
as the host's hormonal and immune systems, that may either restrict or
enhance neoplastic conversion or development.

What emerges here is a picture in which the carcinogenic process is
influenced by a fairly long list of factors, and is either aided or inhibited
by those factors. When we administer a chemical to lab animals and
count tumors at the ends of their lives, we are observing only two
points, and not particularly interesting points, connected by a long
sequence of molecular and cellular events.

We also might be accused of creating a highly artificial situation,
because the cells of genetically homogeneous animals held under strict
laboratory controls do not experience nearly the number and type of
host and environmental influences experienced by people exposed to
the same chemical. Perhaps the most that can be said about chemicals
that are carcinogenic in laboratory settings is that they may, under
some conditions, increase the risk of cancer in humans. Their relative
importance in cancer development depends upon many other factors
not accounted for in the laboratory experiment.

We might now examine some of the information that contributed to
this picture of the carcinogenic process.

Electrophiles

This fancy word is used by chemists to describe organic molecules that
contain groups of atoms that are highly susceptible to reaction with
other groups, called nucleophiles. Nucleophiles are abundant in the
giant DNA molecules. Elizabeth and James Miller and their students at
the McCardle Cancer Research Center, at the University of Wisconsin,
have produced a series of profoundly important scientific papers
showing that many chemical carcinogens are metabolized to compounds having powerful electrophilic properties. The Millers' work
was seminal in establishing that metabolism was essential to the action
of most carcinogens, and also helped reveal the specific chemical
reactions that take place on the DNA molecule.

Metabolism to electrophilic agents is highly important, for example,
in the case of the aromatic amines referred to in the last chapter. These
amines are nucleophilic, not electrophilic in nature, and have little
potential to react with DNA. But they can undergo metabolism to so-
called N-hydroxy derivates. The amine group ($-NH_2$, where N is

nitrogen) is converted to the N-hydroxy group (–N–OH) in cells. The presence of the –OH changes the chemical character of the amine nitrogen atom and, through a sequence of events, the –OH group departs the molecule, taking the electrons that comprise the N–O bond with it, and leaving behind an extremely reactive, and electron-poor nitrogen atom – a highly electrophilic (electron-seeking) group. The electrophilic group sops up electrons from some of the nucleophilic centers of DNA. This reaction between nucleophilic and electrophilic groups can create DNA damage.

The N-hydroxylation reaction does not occur in guinea pigs, and this species does not develop cancers in response to aromatic amine exposure. Rats and most other species do, because they possess the enzymes necessary to create N-hydroxy metabolites. Humans also carry these enzymes, so we would expect them, unfortunately, to respond more like a rat than a guinea pig. This type of information explains a great deal.

Some examples will be seen below of other carcinogens that seem not to require metabolic activation to electrophiles, and which do not possess significant electrophilic properties themselves; these agents appear not to be involved in the initial DNA-damaging event, but rather at later stages of the neoplastic process.

Genotoxicity

Toxicity to the gene is a seminal event in carcinogenesis, if the damage carries through to offspring of the cell that is initially assaulted. A serious omission thus far in this discussion of toxicity concerns the health implications of genotoxic events, of which initiation of carcinogenesis is but one possibility. This omission needs to be corrected before elaborating further on the role of gene toxicity in carcinogenesis.

First, we've talked so far about chemical changes in DNA, and this can certainly be an important form of damage. Chemicals that cause such changes, many of which are electrophiles, are called *mutagens*, and, as we have said, many mutagens are carcinogens. But DNA can be changed in other ways. Its very physical structure can be deranged by certain agents called *clastogens*. In some cases whole chromosomes (the molecular scaffolding of the cell's nucleus that carries the DNA molecules) can be added or lost, a condition called *aneuploidy*. Aneuploidy is often produced by chemicals that interfere with the mechanics of cell division. Not only can genetic damage increase the risk of cancer development, but can be deleterious in other ways. Cell death or abnormalities may occur because some of the protein molecules essential to its existence, and which are created under the direction of DNA, can no longer be produced with fidelity to the cell's original blueprint. If the genetic damage occurs in germ cells, reproductive failure may occur, or if it does not, the abnormal cells are carried forward and may create abnormal offspring. A permanent genetic abnormality may continue in one generation after another; this is a true heritable mutation. Whether environmental chemicals contribute significantly to this type of inherited mutational change is not terribly clear at the present time. Assessing human mutagenic risk for chemicals has been little explored, much less so than mutagenic risks incurred by radiation. The latter have been under study since H.J. Müller's 1927 discovery that radiation could induce mutations in living organisms. Whether congenital diseases such as Down's syndrome (in which an extra copy of a particular chromosome, or aneuploidy, is present) are much influenced by chemical mutagens in the environment remains an active area of study by genetic toxicologists.

One of the more interesting pastimes of the genetic toxicologist is the development of simple tests to detect the capacity of a chemical to produce genetic damage. The challenge is to find the quickest, simplest procedure that is also telling. Many of these tests are performed in glassware (*in vitro*), outside the whole animal. Microorganisms or cells from animals and even from people are placed in liquids containing the nutrients necessary for their growth, and suspect chemicals are added to the liquid. The genetic toxicologists have found a range of clever ways to detect genetic damage in cells grown in this way. They have even found ways not only to test chemicals, but also metabolites of those chemicals; in effect, a means is found to incorporate those enzymes responsible for the mammalian metabolism of the chemicals into the *in vitro* system. Many of these tests can be performed in a matter of hours or a few days at small cost.

In vivo mutagenicity tests are also plentiful, but because they involve whole animals, they are generally more time-consuming (days to several weeks) and expensive.

Professor Bruce Ames, a biochemist at the University of California at Berkeley is one of the pioneers of this type of short-term testing. The Ames Test, as it has been called, is now widely used, typically as one of several short-term tests that constitute a series of tests, or battery. A battery is thought necessary because no single test is adequate to detect all types of genotoxicity. The Ames Test involves the use of mutant strains of a common bacterium, *Salmonella typhimurium*, that 'back-mutate' to their normal states in the presence of a mutagenic chemical or metabolite. Many other bacterial and mammalian cell systems have been made available for this type of testing.

Not too long ago cancer specialists were excited about the prospect of using some of these short-term tests to detect carcinogens. They could replace the very expensive and time-consuming animal bioassay reviewed in the last chapter. After all, these tests detected genotoxicity, an initiating event in carcinogenesis. Genotoxic agents ought to be carcinogens, and those with no genotoxic activity should not be.

Well, this was too simple. Perhaps a high proportion of genotoxic agents are carcinogens, but toxicologists have learned that many chemicals having little or no gene-damaging power are also carcinogenic. Using genotoxicity as the sole criterion for detecting carcinogens would result in missing a number of possibly important agents (although, as shall be seen shortly, it may be that the genotoxic carcinogens are riskier at very low doses than those that act at later stages of the carcinogenic process and that are not genotoxic). Carcinogens that act not as initiators of carcinogenicity, but at later stages of the process, apparently do so through mechanisms not involving gene damage. Gary Williams of the American Health Foundation introduced the categories of genotoxic and epigenetic ('epi' meaning 'outside of') carcinogens in 1977. 'Initiators' and 'promoters' of the process might be other terms for these categories. These general categories are widely acknowledged, although most recognize that a dual categorization may be too simple, and may obscure some important distinctions. Still, the general notion that some carcinogens act at 'early stages' and others at 'later stages,' and that some carcinogens may in fact act at both early and later stages, is highly important and certainly fits well within the multistage model. In fact, the model was constructed upon experimental observations that carcinogens could indeed act in different ways, at different steps of the process. Not all carcinogens are the same, not by any means.

Promotion

Peyton Rous in his laboratories at the Rockefeller University, and Isaac Berenblum of the Weitzman Institute in Israel, together with Philipe Shubik were, back in the 1940s, the first to reveal that certain chemicals, apparently not carcinogenic themselves, could somehow greatly enhance the effects of other substances known to be carcinogenic. The classic experiment involves application of a highly carcinogenic polycyclic aromatic hydrocarbon (PAH, as mentioned in Chapter 7) to the skin of shaved mice, followed by application of substances called phorbol esters. Phorbol esters are complex organic compounds found in croton oil, a natural extract from the seeds of the croton plant. The PAH can, when applied in sufficient amounts, produce excess skin tumors in great abundance. This might be called local carcinogenicity – the mouse skin bioassay has great utility for experimental cancer work because it is relatively rapid and the development of neoplasms is easy to monitor, though not many carcinogens produce skin tumors when applied in this fashion. When the PAH dose is dropped to a low level, such that none or only a few skin tumors would be expected, and the phorbol esters are later applied to the area of the skin treated with the PAH, the yield of tumors climbs dramatically. The phorbol esters are themselves carcinogenically inactive on the mouse skin, but *promote* the development of cells *initiated* by the PAH metabolites, which are highly genotoxic.

Now most people are not exposed to PAHs in this way, and no one is exposed to the phorbol esters, which are strictly laboratory chemicals. But this is not the point. The Berenblum–Shubik promotion studies, and hundreds more that have been explored in the past five decades, have contributed substantially to our understanding of the carcinogenic process. Some chemicals appear to possess both initiating and promoting properties – they are 'complete' carcinogens – while others seem primarily to be involved in the promotion stage of the process. Indeed, Berenblum originally proposed the two-stage model of carcinogenesis that has come to dominate current thinking.

Although the molecular and cellular events associated with initiation are, as we have indicated, fairly well understood, promotion is still fairly mysterious. Somehow the presence of the promoter leads to a breakdown in the system of controls that tissues use to restrict the growth of cells that have undergone neoplastic conversion. Molecular biologists have been examining the processes by which cells communicate and interact with each other – their means for regulating one another's behavior. These interactions occur through some portions of the cells' membranes, known as gap junctions, and it appears that some promoters act at the gap junctions to interrupt the intercellular communication essential to controlling the behavior of aberrant cells.

Promoters may be prominent players in several of the most important human cancers. It appears that bile acids, which are the major components of bile, are promoters for cancer of the large intestine. High fat diets greatly increase bile acid flow through the colon. High fiber intakes help eliminate bile acids, and thus reduce the risk of large bowel cancers. The experimental demonstration that bile acids are promoters of colon cancer helps to explain epidemiological observations that increasing fat intake increases the risk of colon cancer.

Promoters may have a significant bearing on several other human cancers, including those of the breast, ovary and prostate. It is also of more than a little interest that many components of the diet in addition to fat and fiber can significantly modify the response of experimental animals to carcinogen exposure, in some cases enhancing it and in others inhibiting it. In fact, simply restricting total caloric intake can reduce tumor yield, especially in those tissues such as the breast, ovary, and endometrium that are under the control of sex hormones. Some of these dietary influences are no doubt promotional in nature, but most are not well understood. It is quite apparent, however, that nutrient and non-nutrient components of the diet have a major influence on cancer rates, as suggested by the Doll–Peto estimates (Table 5), and the experimental work on diet–carcinogen interactions is beginning to reveal why this is so.

Substances such as promoters that interfere with cell-to-cell communication allow cancer cells to proliferate wildly. But cell proliferation can be induced by other means as well. Toxicity or other types of injury to tissues can result in a proliferative response. So can certain natural and synthetic hormones, such as estrogens, cause proliferation of certain tissues, such as the breast. Chronic viral infections may cause cell killing and its consequence is cell proliferation. It appears that sustained chronic proliferation induced in any of these ways, either by agents foreign to the body or some, such as the estrogens, that are natural to it, can increase tumor growth.

As discussed in the last chapter, the MTD used in cancer bioassays may sometimes be sufficiently high to cause toxicity and cell proliferation, putting affected tissues at extra risk of cancer. It is also the case that rapidly proliferating cells, even if they have not been initiated, are at increased risk of conversion to neoplastic cells. They are more prone to the mutational events that are always naturally present. Professor Bruce Ames holds that 'a high percentage of all chemicals, both man-made and natural, will cause cell proliferation at the MTD and increase tumor incidence.' This is a refined way of saying that almost everything will cause cancer if the dose is pushed high enough, to a level sufficient to cause extensive and sustained cell proliferation. Whether this is true remains a subject of intense debate. Until a consensus emerges among scientists, regulatory authorities will likely not ignore cancer results obtained at the MTD.

Some implications

Some important conclusions emerge even from this rudimentary profile of mechanisms. Metabolism is as significant in carcinogenesis as it is in the production of other forms of toxicity, so a thorough evaluation of risk would require knowledge of species' differences in metabolism, and the influence of the size of the dose on metabolic behavior.

Initiating events involve gene damage and this may result in the fixation in the cell's genetic material of a permanent abnormality. This feature of carcinogenesis perhaps makes it different in kind from most other forms of toxicity. Here the chemical insult occurs and the damage it produces remains in cells even if exposure to the insulting chemical ceases! If doses of the genotoxic agent keep piling up, so do the numbers of those permanent changes. This rather frightening picture is made less so when we recall that cells have a tremendous capacity to repair DNA damage before it becomes fixed, so that not every damaging event, in fact perhaps only a tiny fraction of them, actually translates to a mutation, and only a small fraction of mutations will likely occur at sites that are critical for the development of cancer.

One implication of this view of initiation – and an exceedingly important one – is expressed in the 'no-threshold' hypothesis for carcinogens. Any amount of a DNA damaging chemical that reaches its target (the DNA) can increase the probability of converting a cell to a neoplastic state. This does not mean that every such event will cause a neoplastic conversion, but only that the probability, or risk, of that occurrence becomes greater than zero as soon as the effective target-site concentration of the gene-damaging chemical is reached, and that the risk increases with increasing target-site concentration.

Various mathematical models have been developed to explore the implications for dose–response relations of what is known about carcinogenicity mechanisms. Under some sets of assumptions consistent with these mechanisms, the risk of neoplastic transformation at low doses is directly proportional to the target site dose. This is the so-called straight-line, or linear, model of dose–response; a picture of it is presented in the next chapter.

Actually, the notion that human cancers might result from exceedingly small doses arose first in connection with radiation-induced malignancies. In the 1950s E.B. Lewis of the California Institute of Technology proposed, based on studies of leukemia rates among Japanese atomic bomb survivors and cancer rates among radiologists, that cancer risks might exist at all doses greater than zero, and that a linear dose–response relation is to be expected.

A linear, no-threshold model might be appropriate to describe the dose–response relation for a genotoxic carcinogen, but it is less clear that it is suitable for promoters that act through non-genotoxic mechanisms. Some experts contend that sustained, high level dosing is needed to promote carcinogenesis – the dose needs to be sufficiently large to induce a persistent state of cell proliferation or a breakdown in cell-to-cell communication. Until a *threshold dose* for these toxic effects is exceeded, these experts suggest, significant enhancement of the carcinogenic process is unexpected. All this gets fuzzy when we consider that some complete carcinogens possess both initiating and promoting properties. A fuller discussion of these and selected dose response issues is reserved for the next chapter.

Mechanisms of toxic actions are certainly not all of the type that seems to hold for carcinogens. Some toxic agents act by interfering with the cell's capacity to generate and use energy; its basic metabolic arrangements can be disrupted, leaving it either to die or to operate improperly. Promoters of carcinogenesis are not the only type of toxic agent that can interfere with cell-to-cell communication, and thereby impair the health of a tissue. The cell membrane can be damaged in several different ways by certain toxic chemicals, and this can touch off a series of deleterious events. These various mechanisms, unlike those associated with the production of mutations, all seem to require that some minimum dose of the toxic agent or its metabolite reach the cellular or tissue target, in many cases for extended periods of time – *the threshold dose for toxicity needs to be exceeded.* Toxicologists are by no means certain of this, because mechanisms are not yet worked out in sufficient detail. But, as shall be seen in the next chapter, this view guides most current risk assessments.

The study of toxicity mechanisms will continue. Scientists from the basic disciplines of molecular biology, genetics and biochemistry are increasingly becoming involved in the field of toxicology, and this is a highly desirable trend. The practical payoff, which is to translate what the 'molecular toxicologists' are learning about mechanisms into more accurate characterizations of human risk, is not quite around the corner, but it is surely somewhere in the next block.

9

Dose and Response

We need now to look more directly at a phenomenon that has been briefly touched upon at several points in the preceding chapters: dose–response. Almost everyone has personally experienced this phenomenon, at least in a mild way (consider the relationship between number of glasses of wine consumed and resulting intoxication) and what is intuitively obvious to most people is, in fact, as well-documented a principle of toxicology as can be found. For all chemicals there is a range of doses over which no manifestations of toxicity can be identified in exposed individuals, and there is a higher range over which their toxic properties will begin to appear. Moreover, as the dose increases above that at which toxic effects begin to exert themselves, the frequency and seriousness of those effects also increase. Chemicals differ greatly in their dose–response characteristics, indeed the same chemical may display quite different dose–response behavior in different animal species and under differ conditions of testing, but all exhibit this phenomenon.[13]

An evaluation of the dose–response characteristics of chemicals is at the heart of the problem of understanding the health risks they may pose. If for every chemical in the environment we knew the range of 'no-effect' doses and the point at which toxicity begins to appear – the point at which the *threshold* of toxicity is passed – we could then act to prevent exposures from ever reaching the level at which harmful doses are created (the reader may briefly review Chapter 2 for a discussion of

---

[13] There are categories of toxic response in which the common-sense notion of dose–response does not appear to hold. Allergic reactions are specific to individuals who have been sensitized, and the size of the dose experienced by such people does not appear to matter much. Allergenicity is an important phenomenon but will not be covered here.

the relation between exposure and dose). Risk assessors are, however, not very near capable of achieving this goal. In the first place data on toxicity and dose–response are available for only a small fraction of the chemicals to which people are exposed. Most is known about those chemicals manufactured for use as medicines, pesticides, food additives and for other industrial purposes, and least is known about the tens of thousands of chemicals we consume as natural components of our diets. Recall that part of the impetus for the development of the science of toxicology was the concern for the health of workers exposed to the many products created by the massive revolution in synthetic and industrial chemistry, and by the passage of laws that required toxicity testing of many of these same substances prior to their introduction into commerce. There has been no similar impetus for gaining an understanding of the toxicity of natural chemicals.

The amount and quality of toxicity and dose–response data vary greatly among chemicals. For many there is little more available than information on acute exposures, while for some we know just about everything we need to. For most commercially-important chemicals our knowledge base lies somewhere between these two extremes, although on average it is probably closer to the low end of the scale.

Even if we had an adequate set of data on the toxic properties of a chemical and their dose–response relations, we would still be uncertain about how to identify the 'safe' and 'unsafe' dose ranges. This may sound odd, but it is true. There are several reasons why it is true and understanding them is of utmost importance to an understanding of risk assessment. Why what are nowadays considered 'adequate' data on dose–response are nevertheless inadequate for a thorough and accurate risk assessment is the central subject of this chapter.

The first topic for discussion is the character of the dose–response relations toxicologists and epidemiologists are capable of measuring. We then show their limitations and how they are dealt with; here we enter what is perhaps the most scientifically uncertain and controversial component of the risk assessment process. While the problem is immense it cannot be ignored, because to do so, I hope to show, creates an even more serious problem.

We begin with some factual information about dose and response.

Fundamental properties of dose–response relations

We would like to acquire dose–response information under several different conditions.

First, dose–response curves for acute, subchronic, and chronic exposure conditions would be desirable. (A 'curve' is simply the graphical form of dose versus response. Although it is not always curved, and may take many different forms, scientists generally refer to any such graphical form as a curve.) Toxicologists would also like to be able to construct such curves for the various categories of toxicity that need to be investigated through special studies: reproductive and developmental toxicity, carcinogenicity, perhaps certain forms of immune and nervous system effects, and so on. And, of course, the risk assessor would like to know what the curves look like for the species whose health is to be protected, namely, human beings. Having dose–response curves for each of the different exposure routes experienced by people would also be beneficial. If all this information were available, the risk assessor could identify the most serious forms of toxicity associated with a chemical – generally those occurring at lowest doses – and specify, for different durations and routes of human exposure, the doses that should not be exceeded if toxicity is to be avoided.

As has already been discussed at length, toxicity information is acquired primarily through epidemiological investigations and experimental studies in animals. Quantitative information that can be used to construct dose–response curves can readily be acquired in experimental studies, because the doses are known, and so are the responses related to them. Acquiring solid dose information, as has been seen, is often not possible in epidemiology studies; the epidemiologist may be able to identify broad ranges of exposure experienced by the groups of individuals studied and build a rough approximation of the dose–response curve, but acquiring even moderately accurate historical dose information is often impossible. This is not to say that dose–response curves derived from epidemiology studies are useless, but only to note an important limitation on their use.

More needs to be said about the various measure of dose and response. Perhaps the most common measure of dose is that referred to in Chapter 2: weight of the chemical (mg) that enters the body divided by total body weight (kg) on each day of the exposure period(d). So the typical units are mg/kg b.w./day. A further refinement of the dose measure takes into account the fraction of the weight of the chemical that is actually absorbed through the lungs, GI tract, or skin, the so-called absorbed dose, measured in the same units.

But several other measures of dose are in common use and need to be mentioned. For inhaled chemicals toxicologists sometimes describe the dose as the concentration of the chemical in the inhaled air (in, for example, mg chemical per cubic meter (m^3) of air) *multiplied* by the time (in minutes, hours, or days) over which the air is breathed. The resulting units thus might be (mg/m^3 × hours). Of course, the dose received from such an exposure (in mg/kg b.w./d) can be obtained from information on the quantity of the air breathed by a subject of specified body weight during the time over which the chemical is present in that air. If the effect of the chemical is at the contact point, however, it is likely that the (mg/m^3-hour) expression of dose is more useful than the (mg/kg b.w./d) expression (which would be more useful if the chemical is absorbed and produces systemic toxicity).

A highly useful measure of dose, but one that is not always readily obtainable, is the concentration of the chemical in the body. The most readily measured body fluid that provides a useful indication of dose is the blood. Measurements of urine and expired air levels are easily obtainable, but these media contain excretion products whose concentrations are not always easily related to the dose the body's tissues actually received. The concentrations of a chemical or its metabolites actually in target tissues can also be used as the dose measure; while this is the most meaningful measure, it is also the most difficult to obtain.

Sometimes the measure of dose used by the toxicologist is selected simply because it is convenient, while in other cases it is chosen because it is believed to represent the most telling indication of risk. Perhaps the best example of the latter is the case of lead, where, as we showed in Chapter 6, blood levels provide a more revealing indication of the size of the risk than does mg/kg b.w./day of lead that enters the body. We shall not need to deal extensively with the subject of *dosimetry*, as it is called, and this is fortunate because it is far more complicated than has been indicated here. What is important here is to recognize that several different measures may be used.

Toxic *responses* appear in several guises. We shall simplify the topic a bit and deal with just the two that cover most situations. The first category of toxic responses goes under the technical heading of *dichotomous* responses. This is a fancy term for 'yes/no' responses. In a test of acute toxicity an animal either dies ('yes') or does not die ('no'). In a cancer experiment animals either develop tumors ('yes') or do not ('no'). When the toxic response occurs in this form the toxicologist reports the response as an incidence figure: the fraction of animals that exhibit the effect at each of the doses studied.

The once widely used non-nutritive sweetener, saccharin, produces

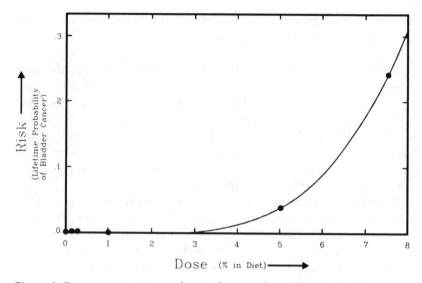

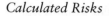

Figure 2. Dose–response curve for saccharin-induced bladder tumors in rats. Dose is expressed as per cent saccharin in the animals' diets. Response (risk) is the fraction of animals in each dosed group developing bladder tumors over their lifetimes – also called the lifetime probability of bladder cancer. Data were reported by Taylor and co-workers, Toxicology and Applied Pharmacology, Vol. 29, page 154 (1974).

excess bladder tumors in rats. The dose–response curve for this effect is presented as Figure 2. The measure of 'dose' in Figure 2 is the percentage of the animals' diets that consisted of saccharin. This is the way the authors presented the dose information; knowledge of the weight of diet the animals consumed each day and the body weights of the animals would allow computation of the dose received in mg/kg b.w./day. The vertical (response) axis represents the fraction of the animals in each of the treatment groups that were found to have developed tumors of the urinary bladder over their lifetimes.

It will be useful to present the results of another carcinogenicity experiment. Saccharin is the least *potent* carcinogen ever detected in an animal study; that is, the dose required to produce a given lifetime incidence of tumors is greater than that of any other known animal carcinogen. Near the other extreme of the potency scale is our old friend, the mold-produced compound, aflatoxin B_1. The dose–response curve for aflatoxin B_1-produced liver tumors in male, Fischer strain rats, is depicted in Figure 3. This result, obtained in the laboratory of Gerald Wogan at MIT and reported in 1974, is the

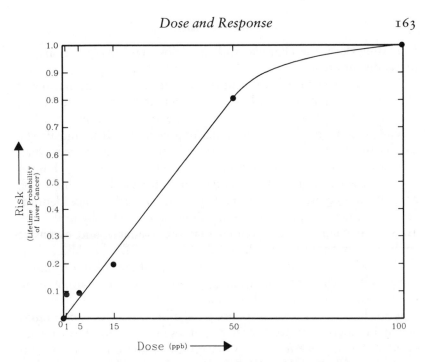

Figure 3. Dose–response curve for aflatoxin-induced liver tumors in rats. Dose
is expressed as parts-per-billion (micrograms per kilogram) aflatoxin in the
animals' diets. Response (risk) is the fraction of animals in each dosed group
developing liver tumors over their lifetimes – also called the lifetime probability of liver cancer. Data were reported by G.N. Wogan and associates in
1974 in Food and Cosmetics Toxicology, Vol. 12, pages 681–5.

strongest response ever seen with aflatoxin – other species of test
animals respond in this way but higher doses are needed.

The first thing to note about the dose–response relation for
aflatoxin-induced carcinogenicity is the fact that a dose only about
one-ten millionth that required by saccharin was needed to increase the
incidence of tumors in the experimental animals. Second, the shape of
the curve is quite different from that of saccharin. In the latter case
several test doses, up to about a dietary level of 3%, produced no
measurable excess of bladder tumors, whereas every dose of aflatoxin
tested, including the lowest, produce an excess of liver tumors. (The
excess tumor incidence at the lowest dose is not statistically significant
in relation to the control, but given the nature of the responses at higher
doses, it would seem appropriate to conclude that the low dose
responses also were induced by aflatoxin.) So, although rodent exposures to both chemicals create increasing tumor incidence with

increasing dose, the characters of their dose–response relations are markedly different.

As a side note, it is worth mentioning that the use of saccharin and aflatoxin as examples is a little deceiving, in that they represent the two extremes of carcinogenic potency in experimental animals. Most carcinogens fall into a narrower range of potencies.[14] It should also be emphasized that, while aflatoxin is inherently more carcinogenic than saccharin, it is not necessarily riskier under actual conditions of human exposure, which were (once) relatively high for saccharin and which are (currently) quite low for the mold product. Toxicity is not identical to risk, we keep emphasizing, and will explain more fully in the risk assessment chapter. Finally, note that the available and extensive epidemiology data on both chemicals do not reveal a causal relationship between exposure to these substances and any form of human cancer, although there is highly suggestive evidence for aflatoxin from several areas of the world where intake is high.

The other type of toxic response we shall discuss is the *continuous* one. In broadest terms we refer to responses in which not the incidence of effect but its *severity* increases with increasing dose. This is best illustrated with an example; let's use the solvent carbon tetrachloride and data from a 90-day toxicity study (Figure 4).

The first manifestation of toxicity is an increase in liver weight at 10 mg/kg b.w./day. This was determined by weighing the livers of control animals and those receiving the chemical. The average group weights were compared, and the statistician determined that the average liver weight for the treated group was significantly (in the statistical sense) elevated over that of the control group. This measure of response is quite different from the 'yes/no' response described in the case of carcinogens. Above the minimum toxic dose – at 33 mg/kg b.w./day, as seen in Figure 4 – the severity of toxic injury becomes increasingly worse (average liver weights increase) and several additional forms of liver toxicity, including cell killing and other manifestations of toxicity, begin to appear.

---

[14] The industrial by-product called 2,3,7,8-tetrachlorodibenzo-*p*-dioxin (2,3,7,8-TCDD), mentioned in Chapter 5, is the only chemical more potent in animals than aflatoxin. This substance, once a contaminant of the herbicide Agent Orange, and a minor by-product of a number of industrial processes, has been the subject of much scientific and public controversy. This story is recounted by Michael Gough in his excellent volume of 1987 called *Agent Orange/Dioxin: The Facts.* (Plenum Press). Whether 2,3,7,8-TCDD, powerful as it is in rodents, has created a significant cancer risk for people is unclear, but the weight of current evidence suggests the answer is no. Unlike aflatoxin, this chemical is not genotoxic, odd for such a potent animal carcinogen. It may act through a threshold mechanism.

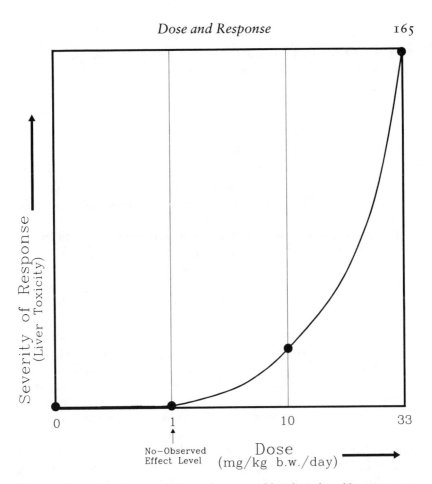

Figure 4. Dose–response curve for carbon tetrachloride-induced liver injury in male rats. The carbon tetrachloride was dissolved in corn oil and doses of 0 (controls), 1, 10, and 33 mg/kg b.w. were administered 5 days/week for 12 weeks. Severe liver injury was observed in all animals at the highest dose, but only relatively mild changes (increased liver weights) were seen at 10 mg/kg b.w. No differences of any type were observed between the control and low dose animals. The dose of 1 mg/kg b.w. is thus a NOEL for this species and this period of dosing. Data were reported by J.V. Bruckner and co-workers in 1986 in Fundamental and Applied Toxicology, *Vol. 6 (No. 1), pages 16–34.*

Continuous and dichotomous responses are both common and, in actuality, both types can be present in a single dose–response measurement. In most cancer studies, for example, both tumor incidence and severity, the latter measured as increased proportions of malignant tumors, increase with increasing dose. For simplicity this fact will be ignored (as, in fact, it is in most evaluations of risk).

Thresholds

The region of the dose–response curve that marks the transition from 'no-toxicity' to 'toxicity' is called the *threshold*. We speak of a threshold dose as that immediately above which responses caused by the chemical begin to manifest themselves.

It seems intuitively obvious that threshold doses exist for all chemicals, but there is considerable debate among toxicologists on this issue. Unfortunately the debate is a tough one to resolve because its resolution requires something like the proof of a negative, which is scientifically impossible. It will be worthwhile to explore this problem in a little detail.

If we refer to alcohol-induced intoxication, the existence of a threshold dose seems obvious. After all, every person can imbibe some amount of an alcoholic beverage without feeling its effects. And we all recognize the point at which the threshold dose is passed. Certainly, individuals vary in their tolerance for alcohol – the size of the threshold dose varies from person-to-person – but there is clearly a 'no-effect', or subthreshold, dose range.

This phenomenon is true not only for alcoholic beverages, but for all chemicals. But we need quickly to point out that so far we are dealing with a limited range of effects, those that can be immediately perceived by the victim. Is it not possible that doses of alcohol below the threshold dose for the immediately perceptible effects of this chemical are causing other biological responses, in the liver for example, that can not be detected by the individual and the ultimate manifestations of which will not become apparent until many years later, in the form of liver cirrhosis? If we think about this a bit we shall quickly recognize that not everyone who drinks alcoholic beverages develops cirrhosis, so there must be a threshold dose for this adverse result of chronic exposure. This seems undeniable, and the only problem is identifying the threshold dose for cirrhosis for a large population of individuals.

Further evidence in favor of a threshold dose for chemical toxicity comes from experimental studies. There are data from thousands of experiments in which doses were identified as having *no observable adverse effect on health*. The maximum dose at which 'no-effects' are observed is called the NOEL: the no-observed effect level. The NOEL, shown in Figure 4 for carbon tetrachloride, is an important data point in assessing risks, as we shall see in the next chapter. Time and again it has been observed that if the dose is lowered sufficiently, no chemical, no matter how toxic at high doses, will be found to produce toxicity. Doesn't this prove threshold doses exist for all chemicals?

Not quite. Remember that data on toxicity are collected in populations of humans or in animal experiments having limited numbers of subjects. In Chapter 6, the problem was described in connection with the discussion of the design of animal experiments. It was shown that in experiments of limited size, there is always a limit to the incidence of disease that can be detected. In typical animal experiments, risks lower than 5–10% can not be statistically measured. To put this in somewhat more straightforward terms, it is always possible that if the numbers of subjects in the animal test, or the numbers of individuals in an epidemiology study are increased, effects might be uncovered that are not observed in studies with smaller numbers of subjects. This is why the modifier 'observed' is inserted in the NOEL: it reflects the fact that scientists can report only that effects were not observed under the specific study conditions, not that they are not producible under other conditions.

It gets worse. Not only are we limited by the size of the study we do, but also by the fact that toxicity tests cannot be performed in all species of animals, nor can it be ensured that the epidemiology study includes individuals exhibiting the greatest sensitivity to a substance's toxic effects. Here we see that the problem of proving that a threshold dose exists is something like the problem of proving a negative proposition – proving that something, in this case a toxic effect, is not produced by a chemical exposure.

To the extent the biological mechanisms underlying the production of toxicity are understood, there would appear to be arguments to be made in favor of thresholds and, as discussed in the last chapter on Mechanisms, others that would cause us to reject the threshold hypothesis for certain classes of toxicity. In favor of the threshold hypothesis are many observations that a certain minimum concentration of a toxic chemical or its metabolites must be present at a cellular target before any biological changes take place. In addition it appears that cells have a remarkable capacity to repair damage caused by foreign chemicals. So until a dose is reached at which repair capacity is simply overwhelmed, there is a very low likelihood that toxic injury will occur. The fact that the body has a wonderful set of mechanisms for bringing about the excretion of foreign chemicals provides additional support for the threshold hypothesis.

Again, however, some counter-arguments can be brought forward. Of particular concern are certain agents, primarily some of the carcinogens and mutagens mentioned in the preceding chapter, that have the capacity to reach the genetic material of cells and cause damage that persists and is passed on from one generation of cells to the next. Cells carrying such damage are at increased risk of developing into the neoplastic state. Recall that not all such cells *will* become cancerous, although there is an increase in the probability (risk) that they will do so. Because it appears that any single molecular event (which does not equate to 'one molecule') may bring about the initial damage to genetic material, then there are reasons to believe that as long as a damaging concentration gets to and chemically alters a critical molecular site in the gene, an increased risk of cancer exists.

Note that this last discussion concerning carcinogens signals a small but significant shift in the notion of a threshold. Under the hypothesis about the mode of action of at least some carcinogens, the genotoxic ones, the concept of threshold is redefined to reflect the fact that we are dealing with increased probabilities, or risks, of cancer occurring. Proponents of the 'no-threshold' hypothesis are not contending that all doses greater than zero 'cause' cancer (though some extremists do). Rather they postulate that all doses above zero *increase the risk* of cancer occurring. This notion of thresholds is different from that represented by the earlier example with alcoholic beverages, where not even a risk of intoxication existed until the threshold dose was passed.

If the dose–response curves for the two sample carcinogens are re-examined, it can be seen that response axes represent the fraction of animals that developed extra tumors at particular sites – the bladder in the case of saccharin, and the liver in that of aflatoxin – over their lifetimes of exposure. Another term for this fraction is *lifetime risk* – the lifetime probability that the specific tumors develop. Each animal in a treatment group was at a certain risk of developing tumors because of the chemical exposure; the size of that risk is represented by the measured tumor rate (ignoring the variability inherent in all experimental studies). At some point, below lifetime risks of about 5–10% (1 in 20 to 1 in 10), the experiments lost capacity to detect excess tumor responses. But under the 'no-threshold hypothesis' risks continue to exist at all doses, even if they can not be detected in animal experiments, and disappear completely only when the dose goes to zero.

Dr. Arthur Upton, formerly Director of the National Cancer Institute and now at the New York University Medical Center, has made important contributions to our understanding of both radiation and chemical carcinogenesis. He summarizes the threshold issue as follows:

Evidence concerning the modes of action of different classes of carcinogens (initiators, promoters, co-carcinogens, and complete carcinogens) suggests that a linear nonthreshold model may be appropriate only for initiating agents and complete carcinogens, whereas models yielding small estimates of risks at low doses might represent more accurately the dose–incidence relationships for other classes of carcinogens. For some types of carcinogens, thresholds might even be envisioned to exist because of relevant pharmacokinetic factors. For example, some chemicals that must be activated metabolically to become carcinogenic may be handled through nonlinear metabolic processes, with the result that thresholds for their carcinogenic effects may exist. In addition, some agents may act through toxic or systemic effects that are produced only at high doses (for example, those causing carcinogenic effects on the mucosa of the urinary bladder in association with cystitis and urinary tract calculi, or those acting through immunosuppressive effects).

If it can be shown, however, that a chemical acts through mechanisms that are shared by agents that contribute to the baseline incidence of 'spontaneously occurring' cancer, then exposure to only a small dose of the chemical can be expected to increase the incidence by some finite amount. For this reason, the use of a nonthreshold model is generally recommended in risk assessment when the mode of action of the carcinogen in question is not known. (references omitted)

Suffice it to say that no clear scientific consensus exists on the subject of thresholds, and may never, because identifying experimental means to resolve the debate is extraordinarily difficult. Perhaps the current state of affairs on this matter can be described as follows:

(1) For all toxic effects except carcinogenesis and perhaps mutagenesis there is probably a threshold dose that must be exceeded before toxicity is created.
(2) For some carcinogens, those that act through genotoxic mechanisms and perhaps some others, there may be some risk at all doses greater than zero (no-threshold hypothesis).
(3) A significant segment of the community of toxicologists holds that at least certain carcinogens do not act through mechanisms that are consistent with a no-threshold hypothesis, and therefore do not create cancer risks until a threshold dose is passed. (See the discussion of promoters, Chapter 8.)

(4) Some toxicologists contend that certain types of noncarcinogenic agents may exert their effects through mechanisms consistent with a no-threshold hypothesis.

The issue may seem arcane, but in fact its resolution has profound implications for the assessment of health risks due to chemicals in the environment. At the present time most risk assessments, at least as performed by regulatory agencies, adopt as working hypotheses the first two assumptions about thresholds or their absence, with the modification that the word 'all' is substituted for 'some' in the first phrase of assumption 2, at least in the United States. But such adoption does not resolve the scientific debate, and many scientists continue to offer evidence and arguments contrary to them. Here we see the first signs of real trouble in the attempt to understand chemical risks. But don't relax too soon, there's much more to come.

Limitations in dose–response data

In addition to the problem of knowing whether a true threshold exists in the dose–response curve, and just where it is, there are at least two other quandries:

The problem of interspecies extrapolation: most dose–response data are available from animals, not people.

The problem of high-to-low dose extrapolation: in most cases meaningful dose–response data can be collected only under conditions of exposure very much more intense than those to which most people are exposed.

These are not the only difficulties, but they are certainly the two that create the most serious problems for the risk assessor.

First this word *extrapolation*. It is, unfortunately but necessarily, used frequently in risk assessment. In essence it describes the process whereby conclusions are reached about one state-of-affairs that either can not be or has not been empirically described, based on information concerning another state-of-affairs that has been subject to empirical study. Scientists speak, for example, about extrapolating toxicity results obtained in rat experiments to reach conclusions about possible toxicity in human beings.

The most reliable forms of extrapolation are those that have been subjected to at least limited empirical validation. Thus, for example, we observe, as we did in Chapter 7, that for almost every substance identified as a human carcinogen there is at least one study showing the substance to be carcinogenic in rodents. This by no means establishes that every identified animal carcinogen will also be carcinogenic in humans. But it does provide some empirical support for making extrapolations from animal data in this fashion. As we also observed in Chapter 7, the animal data are not as telling with respect to the type of cancer that will be produced in humans, so that we have only a very weak basis for extrapolating from animal results to predict the specific site of tumor formation in humans.

The extrapolations from mouse-to-man described in earlier chapters are primarily qualitative in nature, not quantitative. In the context of dose–response evaluations, quantitative extrapolation might concern, for example, estimation of the size of the minimum toxic dose in humans based on observations of the size of that dose in rodents or monkeys. This is trickier, by far, than the type of qualitative extrapolation involved in limited statements such as 'observations of nervous system toxicity in Fischer strain rats are applicable to human beings.'

The twin problems of Interspecies and High-to-Low Dose Extrapolations each has several sets of associated issues, so we shall deal with them one at a time, in as simple a way as possible.

Interspecies extrapolation

In both Chapters 6 and 7 evidence was offered to support the general contention that, with important limitations and exceptions, there are both biological and empirical reasons to adopt the proposition that toxicity results observed in experimental animals are reasonably likely to apply to human beings. This conclusion applies in the qualitative sense only. We have not yet explored quantitative extrapolation issues. Do we expect dose–response relations for humans to match those observed in experimental animals? Are threshold doses for humans equal to those observed in rodents?

Before we can get very far on these matters we need to refine the questions. First, when we speak of dose–response results from experimental studies it is necessary to be specific. For a particular chemical, there may be available, for example, dose–response data from two species, typically mice and rats, and sometimes from others as well. Data may exist for more than one strain of a rodent species and from both sexes. The data may not be of equal quality from all studies and, more often than not, dose–response curves will vary among different species, strains and sexes. Which of these applies to humans?

More complexities arise. No chemical can be tested in all species and strains of mammals. How can we know that a different dose–response curve might not be found if we were to conduct just one more test? We might even find types of toxic effects not picked up in any available tests.

The most puzzling issue of all concerns the 'human being' to whom we intend to extrapolate. If the issue of interspecies extrapolation concerns qualitative inferences only, then it is safe to talk generically about 'human beings', because it is likely that most people will respond, at some dose, to the toxic effects of a substance. But much experience tells us that the dose at which people respond varies among them; some people are much more sensitive than others and will exhibit responses to the same chemical at lower doses. To make matters worse, people who are most sensitive to the effects of one chemical may not be among the most sensitive responders to another chemical that exerts its effects by a different mechanism. Discussions of dose–response extrapolation to humans need to take into account the variability in sensitivity among members of the human population.

It is not difficult to understand the wider variability in responses among humans than is typically seen in animals. In experimental studies animals of highly uniform genetic background are held under near uniform conditions and are fed the very same diet day after day. The air they breath and the water they drink are carefully monitored and controlled, as are the temperature, humidity, and even lighting under which they are kept. They are all of the same age and their health status is nearly uniform. All of these environmental and genetic factors must be controlled if an interpretable experimental outcome is to be realized. It is not surprising that a chemical's absorption, distribution, metabolism, and excretion pattern and its rate and mode of interaction with target cells – all of which are determinants of its toxicity – will be fairly uniform among the experimental animals.

The human situation is obviously dramatically different. Large populations are composed of people of diverse genetic backgrounds whose environments, personal habits, ages and health status are almost unimaginably variable. On purely theoretical grounds our expectation is that the dose–response pattern for a large human population will be substantially different from that obtained in animal experiments. Some people are likely more and some less sensitive than experimental animals, although it is rarely possible to identify how sensitivities vary in a large human population.

The expectation of variability in response among humans is not based entirely on this type of theoretical reasoning, but is borne out by empirical investigations of both direct and indirect sorts. The case of lead described in Chapter 6 revealed the increased sensitivity of children relative to adults, both in their uptake of lead from the environment and in the responses of their nervous systems to the absorbed metal. Studies on the effects of air pollutants such as CO, nitrogen and sulfur oxides, and ozone, also described in Chapter 6, clearly show variability in response among individuals, with both health status and smoking habits having profound influences on sensitivity.

It is known from hundreds of animal experiments that dietary changes can drastically shift dose–response patterns for the same chemical in the same species and strain of test animal. Shifts vary depending upon the chemical and its target. It is not unexpected (although there is not a great deal of direct evidence to demonstrate it) that human responses to environmental chemicals will also be diet-dependent. Consider, then, the effect of the highly variable human diet on toxicity.

Experimental studies of many types demonstrate and explain some types of the variability in human responses. The carcinogenic poly-cyclic aromatic hydrocarbons (PAHs) mentioned in Chapter 7 have been very intensively investigated by several research teams interested in uncovering their mechanisms of action. It is quite clear that metab-olism that introduces oxygen atoms at specific locations of the PAH molecules is a critical determinant of carcinogenicity. In an elegant series of investigations Curtis Harris and his associates at the National Cancer Institute found that the inherent activities of the liver enzymes responsible for PAH metabolism varied by up to 100 times among liver cells taken from different people. This kind of variability, which is genetically determined, both predicts and helps to explain differences among individuals in responsiveness to PAH-induced carcinogenicity.

We could go on to offer many illustrations of quantitative differ-ences in the human population in response to toxic insults, but there is no need to belabor the point. Suffice it to say that, in most cases, threshold doses for humans will vary and, for many chemicals the threshold dose for at least some members of the human population will likely be lower than that observed in experimental animals.

How the problems of selecting the appropriate animal result for dose–response extrapolation across species and of estimating the threshold dose for a broad human population are resolved in the risk assessment process will be explored in the next chapter. What has been covered here sets the stage for that discussion.

High-to-low dose extrapolation

For threshold effects the high-to-low dose extrapolation problem is
solved if the threshold dose for the human population can be identified.
If the threshold dose is known for a particular chemical, and the dose of
that chemical received by individuals from whatever environmental
sources create their exposure is also known, then it will be possible to
understand whether those individuals are at risk (whether the environmental dose they receive exceeds the threshold dose) and the extent of
that risk (the fraction of the population experiencing a dose exceeding
the threshold). The problem for threshold toxicants, to be dealt with in
the chapter on risk assessment, is somewhat more complex than this,
but not greatly so.

For those forms of toxicity for which the threshold hypothesis may
not hold – the important case being carcinogenicity – the problem of
high-to-low dose extrapolation is more bewitching. Recall that in
typical animal cancer studies, and in epidemiology investigations as
well, excess lifetime cancer risks lower than about 10% (1/10) are
undetectable. What dose–response relationship holds below the region
of direct observation? And why should we care?

We care because, for almost all carcinogens in the environment,
human exposures and resulting doses are at levels very much lower
than those that create the levels of risk that are sufficiently large to be
detected epidemiologically or in animal tests. We may not much care
that aflatoxin increases the risk of liver tumors in rats, but we surely
care whether people exposed to very much lower doses through certain
foods are also at increased risk of cancer. We care that benzene has
been shown to increase leukemia risk in certain relatively small groups
of highly exposed workers, but we perhaps care more whether the
millions and millions of people daily exposed to levels in the air many
thousands of times lower are also at increased risk of leukemia. We
would also like to know the sizes of these risks.

Well, unfortunately, caring to know the answers to these and
hundreds more such questions does not allow us to know. As we have
noted several times, no means are now available to obtain direct
measures of risks below the 5–10% range, so if we seek empirical data
to answer these questions we shall be disappointed.

We can, however, set up some hypotheses about the nature of the
dose–response curve in the region between zero risk and 10% risk. To
construct our hypothetical dose–response curve we should attempt to
ensure that its nature is at least consistent with whatever current

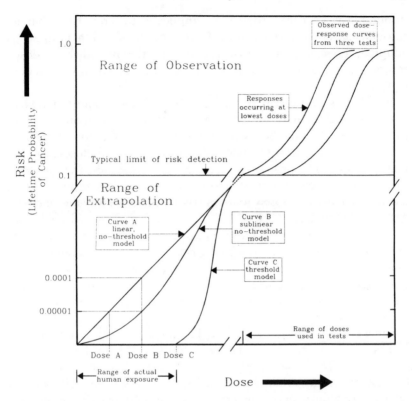

Figure 5. Hypothetical dose–response curves for chemically-induced carcino-genicity, showing measured dose–response curves from three studies (top right-hand quadrant) and some possible ways those curves might behave in the low dose–low risk region (lower left-hand quadrant, in The Range of Extrapo-lation). Note: The graph is not drawn to scale. The lower left-hand quadrant has been greatly expanded to show the possibilities for extrapolation. See text, pages 174 to 179 for a full discussion of this graph.

understanding biologists possess regarding the mechanisms of carcino-genesis. It is also important to ensure that it is not inconsistent with any known data regarding carcinogen behavior. A set of hypothetical dose–response curves are set out in Figure 5 to assist the following discussion.

In the discussion of thresholds presented earlier we pointed out that many scientists adopt the no-threshold hypothesis, if not for all, then for at least those carcinogens that act through genotoxic mechanisms. Some biological basis exists for such an assumption, but it is also disputable as a general proposition, applicable to carcinogens that act through different mechanisms. If the no-threshold hypothesis is adopted, however, one critical point on the hypothetical dose–response curve is automatically fixed: risk equals zero only when dose equals zero.

Well that is a little help, but there are many more points to fix. Here we enter the domain of the 'linear vs. non-linear.' Based on one set of theories regarding the behavior and mechanism of action of carcinogens, a so-called *linear* model of the hypothetical dose–response relationship can be proposed (Figure 5, Curve A); these theories have their origin in research on radiation-induced cancers and on certain chemicals that seem closely to mimic radiation in the way they increase cancer risk – that is, by initiating gene damage, as reviewed in Chapter 8. The linear model, at least for very low doses, is also supported by the fact that there is a substantial background of mutagenic events taking place all the time; it is mathematically rigorous that any addition to that background, as long as the mechanism of mutation is the same, will increase risk in direct proportion to target site dose – the dose–response is linear.

Some experts believe the so-called sublinear (or curvilinear) model (Curve B in Figure 5) is a better description of the low-dose risk relationship. This model displays the no-threshold property, but shows less risk at low dose than does the linear model. It is favored by those who worry about genotoxic agents, but who point to the biological mechanisms that reduce the effectiveness of these agents at low doses. Others, of course, opt for the threshold model (Curve C in Figure 5) on the grounds that all organisms can tolerate a certain amount of exposure to any toxic agent, including carcinogens, especially those that promote or act only at later stages of the process.

An important scientific paper on this topic appeared in 1961 in the *Journal of the National Cancer Institute*. It was authored by Nathan Mantel, an eminent biostatistician, and W.D. Bryan. The paper was entitled 'Safety testing of carcinogenic agents,' and offered a methodology for extrapolating cancer bioassay data to low doses. The Mantel–Bryan method did not involve a linear model, but rather one based on the statistical concept of tolerance distribution. This mathematical model includes the assumption that individuals in a population can tolerate a certain dose without incurring a significant risk, and that tolerances vary among individuals. In fact, they vary in a particular, definable way (what statisticians call a log-normal distribution). Mantel and Bryan showed how the model could be used to estimate upper limits on low dose risks. By referring to a fairly large set of dose–response data, they were able to fix some of the model's parameters to ensure low dose risks would not be underestimated, although of course this could not be proved unequivocally. They also defined a 'virtually safe dose' as one that would be highly unlikely to create an excess lifetime cancer risk greater than the exceedingly low probability of *one in one hundred million*! The Mantel–Bryan risk extrapolation method has fallen out of favor, at least in regulatory circles, but what remains in place is their proposal that 'safe doses' could be defined for agents that appeared to act through non-threshold mechanisms.

It is important to note that Mantel and Bryan never offered their method as one that could predict the actual risk associated with a given, low dose. They rather saw the procedure as one that could be used to identify the dose that would be unlikely to create a risk greater than the operationally-defined 'safe' level – a subtle but important point. These authors had no greater ambition for their method.

During the 1960s and 1970s, and continuing to this day, a handful of biostatisticans continued to develop the Mantel–Bryan idea, not with respect to the issue of the safety of any particular level of risk, but with respect to the mathematics of the dose–response relationship. Some raised doubts regarding the mathematics and certain other features of the Mantel–Bryan procedure. Some offered alternative low-dose extrapolation models. In no particular order, Kenny Crump, David Hoel, Marvin Schneiderman, Jerome Cornfield, John Van Ryzin, Daniel Krewski, Charles Brown, Alice Whittemore, David Gaylor, Ralph Kodell, Robert Sielken, H.O. Hartley, and Thomas Starr, and just a few others, working at universities, in government agencies, and in industrial firms, have been the leading explorers of the biological and statistical bases for low-dose extrapolation. Some have built upon the earlier, fundamental work of giants such as Peter Armitage and Sir Richard Doll, who had worked out a mathematical description of the multistage model of carcinogenesis in the early 1960s. Some showed how the Armitage–Doll model could be practically applied, and taught why it supported a mathematical form that was linear at low doses. Others worked out alternative mathematical models, based either on more purely statistical ideas, or, in some cases, upon somewhat different biological notions. Modifications of the linear model necessary to incorporate ADME data have been much explored by still others. A few have been particularly successful at showing how dose–response data from epidemiology studies could be modelled. These scientists have built upon each others' work, and have offered up an array of mathematical models for low-dose extrapolation, some of which are used by regulatory agencies, and others of which are being applied by others in the risk assessment community. But these mathematically-oriented biologists would be first in line to say that none of these models is fully satisfactory or has been empirically verified; they have greater or lesser degrees of utility and biological support, but, as we have emphasized over and over, the true mathematical form of the dose–response relationship at low doses is not known to us, and will not be until we have much better knowledge of the mechanism(s) of carcinogenesis.

The essential characters of these various models though by no means their actual forms, have been summarized in graphic form in Figure 5. This representation does not do justice to the mathematics of the process, but is adequate to portray what is going on when risk assessors engage in high-to-low dose extrapolation.

Return now to Figure 5. We see that the three extrapolation models depicted (represented by curves A, B, C) predict quite different risks in the low dose region.[15]

Let us specify some very low risk level as that which we would not like to see exceeded. Pick an extra lifetime probability of cancer of 0.00001 (1 in 100 000). For purposes of this discussion, this risk will be considered 'tolerable.' This probability is 10 000 times smaller than the minimum risk for which we have dose–response data (0.1, see Figure 5). If we adopt hypothetical dose–response model A, the linear, no-threshold model, then we see that *dose* A produces an extra lifetime risk of 1/100 000. Similarly, model B yields *dose* B as producing the same risk.

Dose B is greater than dose A. This means that under the sublinear, no-threshold model, individuals could tolerate a dose larger than they could tolerate if the linear model, predicting dose A as the maximum allowable, were correct. To put this the other way round, individuals exposed to dose B would experience an extra lifetime risk of 1/100 000 under the sublinear model, but would experience a higher risk (1/10 000 or 0.0001 in the example of Figure 5) if the linear model were correct. The linear model thus predicts higher risks at a given dose than does the sublinear model, or, conversely, the linear model requires

[15] It turns out that many mathematical models of dose–response that yield quite different results at low doses match observed dose–response data equally well; the degree of 'fit' to observed data doesn't help identify the 'best' model for very low doses.

lower tolerable doses than does the sublinear model at the same level of tolerated risk. Note that in our example, the threshold model (C) predicts no risk at all until dose C is reached.

Although there is no consensus on the matter, most scientists believe that models showing even higher risks than the linear model, so-called superlinear models, are not likely to be correct. They see the linear, no-threshold model as a kind of *upper limit* on the dose–response relationship, and admit that the actual but unknown dose–response relation may be sublinear or may even take the threshold form for some carcinogens.

The fact is scientists do not know the dose–response relation at very low risks for any carcinogen. The various hypothesized relationships have different degrees of scientific support but none has been documented with anything like the degree of rigor usually sought by scientists. Moreover, different low dose extrapolation models can yield markedly different estimates of risk, with variations of 100- or 1000-fold not unusual. Here's another issue left to the unhappy risk assessor to resolve.

10

Assessing Risks

I believe it was Mark Twain who quipped, when asked what he thought of the music of Richard Wagner, that 'It's not as bad as it sounds.' Risk assessment might be similarly described, by both its supporters and its detractors!

Risk assessors, in apparent disregard for the terrible uncertainties associated with interspecies extrapolations and identifying dose–response relationships, have in recent years been making one announcement after another on the risks associated with carcinogens in the environment. 'There will be greater than a one in one hundred thousand chance that ALAR will cause cancer in children consuming apple juice.' 'One person in every ten thousand will contract cancer from residues of ethylene dibromide (the grain fumigant, EDB, used until EPA's ban in 1984) in flour.' 'Dioxin contamination of bleached paper products creates a cancer risk greater than one in 500 000.' The media contain revelations of these types with increasing frequency, usually accompanied by statements from regulatory agencies designed to quell public fears, remarks from manufacturers to the effect that risks have been greatly exaggerated, and professions of outrage from critics of both the regulatory and industrial communities.

I hope it is obvious by now that statements about risk of the type cited above, if standing alone, are simply false. There are no means available to identify these types of risks with the degree of certainty suggested by the language used. Any risk assessor who contends that 'x people will contract cancer,' where x is a single number, is either a liar or is highly incompetent. Even if x is reported as a range of numbers, use of the phrase 'will contract' is dishonest. Perhaps the best a risk assessor might do, given today's knowledge, is a summary that goes something like this:

Difluoromuckone (DFM) has been found to increase the risk of cancer in several studies involving experimental animals. Investigations involving groups of individuals exposed in the past to relatively high levels of DFM have not revealed that the chemical increases cancer risk in humans. Because these human studies could not detect a small increase in risk, and because there is a scientific basis for assuming results from animal experiments are relevant to humans, exposure to low levels of DFM may create an increase in risk of cancer for people. The magnitude of this risk is unknown, but probably does not exceed one in 50 000. This figure is the lifetime chance of developing cancer from a daily exposure to the highest levels of DFM detected in the environment. Average levels, which are more likely to be experienced over the course of a lifetime, suggest a lifetime risk more like one in 200 000. These risk figures were derived using scientific assumptions that are not recognized as plausible by all scientists, but which are consistently used by regulatory scientists when attempting to portray the risks of environmental chemicals. It is quite plausible that actual risks are lower than the ones cited above; higher risks are not likely but cannot be ruled out. Regulators typically seek to reduce risks that exceed a range of one in 100 000 to one in 1 000 000. Note that the lifetime cancer risk we face from all sources of these diseases is about 1 in 5 (1 in 10 for non-smokers), so that, even if correct, the DFM risk is a minor contributor to the overall cancer problem. Prudence may dictate the need for some small degree of risk reduction for DFM in the environment.

This statement could no doubt be much improved upon, but it is certainly much closer to what risk assessors know than the ones cited earlier. The purpose of this chapter is to reveal why this is so, by attending more completely than we have so far to the fundamentals of risk assessment.

Assessing and managing

Look back at the last sentence in the long quote just presented. It is different from all the other sentences in the paragraph because it does not deal with describing the risk. Instead it suggests that some action should be taken to reduce the risk.

The last sentence is outside the bounds of what is called risk assessment. The risk assessor may hold that opinion, but should avoid including it as part of the risk assessment. The latter term refers to the process whereby all available scientific information is brought together to produce a description of the nature and magnitude of the risk associated with exposures to an environmental chemical. The term *risk management* is used to describe the process whereby decisions are made about whether an assessed risk needs to be reduced to protect public health, and on the means that should be used to achieve the desired reduction. Assessing is covered in this chapter, managing in the next.

An emerging discipline?

Like those of toxicology, the foundations of the risk assessment discipline were laid in several different areas of study, and these have begun to merge only within the past decade. One of its foundations can be located in the work of radiation biologists and health physicists who began, not long after the discovery of radioactivity at about the turn of the century, to investigate the adverse health consequences of exposure to this form of energy. Their work received a major impetus from the development, production and deployment of nuclear weapons; all three activities created opportunities for human exposure to various forms of radiation, the most extensive of which were those incurred by survivors of Hiroshima and Nagasaki. Many of the models now used for assessing low dose risks from chemical carcinogens have been in use by radiation scientists for several decades; even those scientists have not been able to resolve the 'threshold vs. no-threshold' and 'linear vs. non-linear' debates.

Another important foundation for risk assessment can be found in the work of safety engineers. For several decades, concerns about the safety of large physical structures and complex manufacturing and energy-production facilities – dams, nuclear power plants, chemical-manufacturing facilities, and so on – have prompted analyses of the risks that they may fail to operate as planned. Failure analysis, as it is sometimes called, involves assigning probabilities to various events that may lead to a failure – the release of a highly toxic chemical to the atmosphere, for example – so that construction and operating procedures, and various 'fail-safe' mechanisms, can be appropriately built into the system. Safety engineers are, of course, also involved in the production of hundreds of types of complex manufactured goods, the failure of which could lead to injury.

The Society for Risk Analysis was organized in 1980 by a group of scientists and engineers from these various disciplines, and toxicologists were included among them. These individuals believed that, whether the issue was the failure of a nuclear power plant or brakes on an automobile, or human exposures to chemicals or to radiation, they were united by a common interest in the analysis of risk. The Society and its journal *Risk Analysis* have prospered and continue to draw new membership. A meeting of The Society brings together an odd but interesting assemblage of engineers, health scientists, statisticians, toxicologists, physicists, molecular biologists, radiation experts, regulators, and even social scientists and psychologists interested in problems of risk perception and communication. It appears that risk analysis is here to stay, and risk analysts stand ready to explore the threats of just about any aspect of modern technology and the natural world.

Risk

Risk is the probability that some harmful event will occur. What is the probability that certain types of cancer will develop in populations exposed to aflatoxin in peanut products or benzene from gasoline? What is the likelihood that workers exposed to lead will develop nervous system disorders?

Because it is a probability, risk is expressed as a fraction, without units. It takes values from 0 (absolute certainty that there is no risk, which can never be shown) to 1.0, where there is absolute certainty that a risk will occur. Values between 0 and 1 represent the probability that a risk will occur.[16] We say that the lifetime cancer risk from carcinogen A at an average daily dose of B is one in 100 000 (0.000 01); if this number is accurate, it means that one of every 100 000 people exposed to carcinogen A at a lifetime average daily dose of B will develop cancer over a lifetime. The probability also describes the extra risk incurred by each individual in that exposed population.

People are more familiar with expressions of risk associated with various activities than they are with risks associated with chemical exposures. We speak, for example, of the annual risks of dying as a result of certain activities. The annual chance of dying in automobile accidents for people who drive the average number of miles is about

---

[16] In Chapter 7, in the section on epidemiology, risks were presented in *relative* terms, the ratio of the risk observed in one population to that observed in another. Such relative risks are important and useful, but differ from the absolute risks discussed here.

one in 4000. The average bicyclist faces an annual risk of death from pedalling of about one in 30 000. Pack-a-day smokers who began at age 15 incur a risk of death from lung cancer of one in 800. The lifetime risk of developing cancer in the United States is about one in 5.

These types of expressions of risk are more familiar to people, but they mean roughly the same thing as those described earlier for the risks of toxicity from chemical exposures – with at least one exceedingly important difference.

Information on death rates from automobile or other types of accidents or activities is generally much more solid than that pertaining to most chemical risks. Statistical data, compiled by actuaries, is used to derive such risk information. There's uncertainty associated with these actuarial figures, but most are fairly reliable.

Most of the risks associated with environmental chemical exposures are much less firmly known. So, although chemical risk information is often expressed in the same form as that based on directly measured risks, they are derived using quite different methods, and almost always include extrapolations beyond measured risk data.

Another important source of confusion in the use of the term risk needs to be brought out here. When a risk assessor states that exposure to DES increases the risk of certain cancers in women, she means that, under certain DES exposure conditions, there occurs a greater number of those types of cancers than the number that occur in the absence of DES exposure. The risk assessor also means that DES has contributed to the cause of those extra cancers, in the sense that its presence directly brought about certain changes (probably related to its endocrine tissue-stimulating properties, a property of all estrogens) that enhance tumor development. A goal of risk assessment is to estimate the extra risk caused by a toxic or carcinogenic chemical over that which exists when exposure to the chemical does not exist.

This notion of risk is not to be confused with what is called a *risk factor*. Physicians say that people who are overweight are at increased risk of heart disease. But this does not necessarily mean that the heart disease is caused by the obesity. Rather, obesity is what is called a *risk factor*: physicians know from much correlational data that heart attacks occur more frequently in individuals who are overweight, but do not have compelling evidence that it is the extra weight that is the cause of those heart attacks. Other factors, correlated with both obesity and heart attack rates, are probably the underlying causes. Knowing risk factors is exceedingly important, because physicians can use this information in treating diseases. But it is important to keep in mind the difference between a risk factor and risk, which carries with it the connotation of a true contribution to causation.

Assessment

Risk assessment might be viewed simply as a means of organizing and analyzing all available scientific information that bears on the question at hand. If we are interested in understanding the nature and size of the health risk associated with, for example, aflatoxin in peanut products or trichloroethylene in drinking water, there are three types of information that must be evaluated.

The first type concerns the *hazards* of these chemicals: their flammability, their explosivity, their radioactivity, and their toxicity. In the present context we are interested in the toxic properties of these chemicals. So, it would be necessary, under what is called the *hazard evaluation* step of risk assessment, to assemble all the available epidemiology and experimental toxicity data (the latter to include animal toxicity studies, ADME data, and studies of mechanisms of toxic action). The assembled data would then be critically evaluated to answer the question: what forms of toxicity can be caused by the chemical of interest, and how certain can we be that human beings will be vulnerable to these toxic effects (under some conditions)?

Because we know that the nature, severity, and risk of toxicity vary with dose, the next step in a risk assessment is the *dose–response* evaluation. For each of the established forms of toxicity caused by the chemical of interest, what is the quantitative relationship between dose and risk of toxicity in the range of doses that have been or might be experienced by human beings?

And what is that dose range? *Human exposure evaluation* comes next. What populations are of interest? For aflatoxin it would be all individuals who consume peanut products. For trichloroethylene it might be those individuals who consume water derived from contaminated ground water supplies. What dose of the chemicals do these individuals receive, and for what period of time? Because not all individuals in the population groups of interest will be exposed to identical doses, the risk assessor would attempt to understand the distribution of doses in the populations: the number of people exposed to each of several different doses, or dose ranges.

These three steps – hazard evaluation, dose–response evaluation, and human exposure evaluation – provide all that is necessary to answer the ultimate risk question: what type of toxicity is expected in the exposed population (neurotoxicity?, birth defects?, cancer?) and what is the risk (probability) of it occurring in that population? The answer to the last question provides the risk assessor's final *characterization of the risk*. The long quote about the made-up chemical DFM at the start of this chapter is a summary risk characterization.

The four steps of a risk assessment were first described and recommended as recently as 1983 by a committee of the National Research Council (NRC). Although risk assessors had for a number of years been conducting these same types of evaluations, the NRC committee's recommendations were important because they served to create a common language among risk assessors and encouraged the development of consistency in the assemblage, organization, and evaluation of the information base upon which risk assessments are built. Nowadays a near-consensus on this approach to organizing risk information exists in the scientific community concerned with risks of chemical toxicity. In fact, it would seem that almost any type of risk could be effectively analyzed within this same framework.

The discerning reader may have noticed that the contents of the chapters of this book match the steps of risk assessment. Hazard evaluation concerns the type of material presented in Chapters 3 through 8. Dose–response was treated in 9, and human exposure evaluation was briefly reviewed in Chapters 1 and 2.

While the scientific basis for risk assessment has been described in the previous chapters, certain critical issues were left up in the air. Scientists have not yet been able, for example, to establish with certainty the relevance to humans of animal toxicity findings, and generally have poor or even no empirical data regarding dose–response relations at the human dose levels typically associated with environmental chemicals. Moreover, toxicity data gaps of one sort or another exist for all chemicals; sometimes the gaps are not of great importance, but often they will seriously hamper the risk assessment process. So, while risk assessors may know much, they are always faced with significant scientific uncertainties. Indeed, some observers think risk assessment is best described as the analysis of uncertainty.

Uncertainty and science policy

No one likes to deal with uncertainty in any aspect of life, let alone those aspects having to do with possible threats to health. At the same time the scientific uncertainties regarding the health risks posed by chemicals in the environment are undeniable. It will be seen that risk assessors have a technique for dealing with these uncertainties, but the technique does not make them vanish. The risk assessor's technique simply attempts to ensure that uncertainties are treated in the same way for all chemicals. The risk assessor can claim, at best, that the *relative* risks posed by different chemicals can be roughly understood, (and this is exceedingly useful information), but is prohibited from making anything but weak statements regarding the absolute risk posed by a specific chemical. (Keep in mind that these remarks apply to risks of toxicity associated with chronic, low dose environmental exposure to most chemicals; there are some chemicals and types of exposures – particularly acute ones – whose risks are quite well understood).

The principal authorities under which risk assessments are conducted are the various federal and state regulatory and public health agencies. These authorities have, over the past several decades, adopted certain policies regarding the conduct of risk assessments to serve regulatory decision-making. These policies primarily concern the specific assumptions that are to be used in the conduct of risk assessments to deal with gaps in basic knowledge or in the data available on specific chemicals. In Chapters 8 and 9, for example, several scientifically plausible models were shown to exist for describing dose–response relationships for carcinogens in the range of doses and responses below the observable range; they were summarized in Figure 5. Scientists can not be sure which is correct; the best they can do is set forth the degree of support, in no case anywhere near complete, that each can claim. As revealed in Figure 5, the various models yield sometimes substantially different pictures of the risk for the same exposure. If a risk assessment is to be completed, a *science policy choice* (the phrase used by the NRC) must be made about the model to be used for low-dose extrapolation for carcinogens. Several similar choices having to do with other uncertainties are needed to complete most risk assessments.

Regulatory agencies make science policy choices. These agencies have a duty to provide some answer to questions regarding environmental sources of risk, and then to act on the basis of those answers. Science will take the agencies a long way toward an understanding of risk, indeed just about all the way for some chemicals, but assumptions – science policy assumptions – are needed to arrive at an answer for most chemicals.

What can be said about an 'answer' so arrived at? How much truth can be claimed for it? Is it reasonable to act upon it, for example by calling for sometimes very costly and socially disruptive measures to reduce the risk?

No unarguable answers to these questions can be found. But it might help if we examine the science policy choices made by regulatory agencies.

Risk assessment in U.S. regulatory agencies

Although exceptions exist for almost all of them, a set of general principles under which regulatory agencies in the United States conduct risk assessments has evolved over the past several decades (those concerning carcinogenic risks are of more recent origin). The important ones might be stated as follows:

Principle (1) In general, data from studies in humans are preferred to animal data for purposes of hazard and dose–response evaluation.

Principle (2) In the absence of human data, or when the available human data are insufficiently quantitative or are insufficiently sensitive to rule out risks, animal data will be used for hazard and dose–response evaluation.

Principle (3) In the absence of information to demonstrate that such a selection is incorrect, data from the animal species, strain, and sex showing the greatest sensitivity to a chemical's toxic properties will be selected as the basis for human risk assessment.

Principle (4) Animal toxicity data collected by the same route of exposure as that experienced by humans are preferred for risk assessment, but if the toxic effect is a systemic one, then data from other routes can be used.

Principle (5) For all toxic effects other than carcinogenicity, a threshold in the dose–response curve is assumed. The lowest NOEL from all available studies is assumed to be the threshold for the groups of subjects (humans or animals) in which toxicity data were collected.

Principle (6) The threshold for the human population is estimated by dividing the NOEL by a safety factor, the size of which depends upon the nature and quality of the toxicity data and the characteristics of the human population. The estimated human threshold dose has several different names, depending upon the regulatory context (see text below).

Principle (7) For carcinogens a linear, no-threshold dose–response model is assumed to apply at low doses, as in Figure 5, Curve A.

Principle (8) Generally, human exposures and resulting doses and risks are estimated for those members of the population experiencing the highest intensity and rate of contact with the chemical, although other, less exposed subgroups and people experiencing average exposures will frequently be included.

Other principles related to other aspects of the regulatory risk assessment process are important, but these eight are the major ones.

The United States' regulatory position on carcinogens,[17] which assumes all to be of the riskiest kind (Principle 7) seems odd, given all we have read about the important mechanistic distinctions among toxicants of this class, and the possibility of sublinear or even thresholds in the dose–response curves of some of them. Because mechanisms are not fully understood, and because it is not always clear whether a specific carcinogen is of one variety or the other, or has some of the characteristics of both, regulators have made the science policy decision to remain cautious. They leave open the possibility that they can be proved wrong, but it is unclear how far toward 'certainty' research scientists will have to travel before regulators see fit to incorporate into risk assessments the knowledge they acquire about mechanisms.

Note that all these principles tend to be 'health-protective' in nature. That is, where science is uncertain, and where a range of plausible choices exist, regulatory policy favors that which yields the highest estimate of risk. In addition to Principle 7, the clearest examples are Principles 3, 5, and 8. If animal toxicity data on a chemical are available from several species, and they differ in dose–response characteristics, regulatory policy favors the species showing the greatest response. This principle is illustrated in Figure 5, where it is seen that the high-to-low dose extrapolation was based on one of three observed dose–response curves – specifically, that representing toxic responses occurring at the lowest doses. It may well be that the response seen in a less sensitive species is 'more relevant' to humans, but unless data are available to show the greater relevance, then the data from the most sensitive species will be used to determine potential human risk. If ADME data were available, for example, and showed that the less

---

[17] The risk models described herein for carcinogens are not yet much used outside the United States. Carcinogens tend either simply to be banned, where it is possible to do so, or are treated the way non-carcinogens are treated in the United States. The discussion of risk assessment as applied to non-cancer forms of toxicity is thus applicable to carcinogens in countries outside the U.S.A.

sensitive species was more like human beings in the way it metabolized the chemical, then this might justify selection of that species for risk assessment. Of course, the ADME data would have to be quite clear and complete to support such an approach. And in the absence of any such compelling ADME data, the most sensitive species would be chosen as the human surrogate. Similar 'health protective' choices are reflected in the other principles – recall, for example, the discussion in Chapter 9 on the linear, no-threshold model for carcinogens as representing a plausible upper limit on risk.

A particularly important component of the risk assessment concerns the assumptions and data used to develop an estimate of human exposure – the human dose. Elements of the dosimetry problem were described in Chapter 2 and we do not intend to go into them in more detail here. Certain additional aspects of this issue should be brought out, however, because they critically affect the meaning that is to be attached to the risk estimates produced under the regulatory principles.

Regulators, and other risk assessors as well, frequently engage in something close to what is called a 'worst-case' exposure analysis. Here are three simple examples.

A petroleum refinery emits benzene. Assume a hypothetical resident lives at the plant's fenceline for a full lifetime and spends most of every day at home. Calculate the daily benzene dose this hypothetical person receives, and then estimate his cancer risk.

Assume a person consumes food at a very high rate (90th percentile of consumers of that food), and that every mouthful for a whole lifetime contains a pesticide residue at the maximum allowed concentration. Calculate this person's dose and risk.

Assume a worker is exposed to the maximum allowable air concentration of a workplace carcinogen, eight hours every day, five days every week, for a working lifetime of 40 years. Calculate the worker's dose and resulting risk.

Just what do risk estimates based on such assumptions mean? To varying degrees they are all fairly improbable, 'worst-case' exposure scenarios. They are, however, easy to work with – doses can be estimated under these assumptions with very little real information and the calculations can be done by a high school chemistry student. Additionally, if risks estimated under such assumptions are very low, we can be doubly assured that actual risks are probably not significant. Of course, if they are high, as they often are, one wonders what to make of them. The best course would be to return to the drawing boards and to try to obtain some real data to substitute for the assumptions used. In any case, risk estimates developed under such assumptions, it should always be emphasized, probably apply to few real persons.

Regulatory agencies also attempt to develop more realistic estimates, but this is difficult, and a scientific consensus on just what exposure pattern should be presumed desirable for risk assessment is not available, except for a few circumstances (food additives, human drugs, and a few others).

For these several reasons and for some additional statistical ones we have avoided, regulatory risk estimates, derived under the Eight Principles, are typically characterized as *upper limits* on human risk. Because the critical science policy assumptions tend to err on the side of safety, the resulting risk estimate may be about equal to or, more likely, will be greater than the actual risk. Some critics argue that because conservative assumptions are sometimes piled one upon the other in the regulatory risk assessment process, then the resulting estimates are highly improbable – they are likely to be greatly in excess of any actual risk. This is probably correct for some regulatory assessments, but the only way to judge this is to inquire deeply into all of the data and assumptions that underlie the assessment.

As in the case of the ADME data discussed above, regulatory policy permits departures from the standard principles in particular cases, but only when data are available to allow such departures to be made without creating a significant possibility that risk will be underestimated. Experimental data that reveal the mechanisms of action by which a specific chemical produces carcinogenicity might show, for example, that a sublinear or even a threshold model is to be preferred to a linear model, at least for that one chemical. One type of data that is beginning to influence regulatory science policy is that resulting from what are called *pharmacokinetic* studies. The excellent experimental work of investigators such as Richard Reitz at Dow Chemical and Melvin Andersen, now of the Chemical Industry Institute of Toxicology in North Carolina, has inspired enormous interest in this topic. The term 'pharmacokinetics' was first used in connection with pharmacologic agents, but is now also used for chemicals that produce adverse effects. It is a fancy term that applies to ADME studies, but with emphasis on the *quantitative measurement of rates* of absorption, distribution among tissues, excretion and metabolic changes. Here's how such data may cast some light on dose–response relations.

Consider a chemical that increases cancer risk at specific body sites in laboratory animals, over a range of doses. Although it was not emphasized under the discussion of dose–response modelling, one of the biological underpinnings of those models is that risk is a direct function of the concentration of carcinogenic metabolites at the critical target site, typically the genes of the affected cells. The risk assessment is based, however, not on target (cellular) doses, but rather on the doses administered to the animals. The reason for this is that target site doses are not usually measured in cancer tests. The dose–response models include the assumption that the critical measure of risk, the target site dose, is in some constant relationship to the administered dose at all dose levels. For example, assume that at dose 1, for every 100 molecules of a chemical that enter the body, 10 molecules of its carcinogenic metabolite reach target sites (the other 90 follow a different path or are metabolized differently). At dose 2, for every 100 molecules that enter the body, 10 molecules of the damaging metabolite reach target sites in the form of metabolite. And even at dose 3, ten per cent of the molecules entering the body reach target sites. So, while target site doses are only a fraction of the administered doses, that same fraction (10%) holds at all administered doses. If this is true then, under the linear model, the administered doses can be used as the measure of risk, because target site doses are, at all doses, in direct proportion to administered dose.

But suppose that, in reality, the constant fraction relation does not hold. Suppose that at dose 1, for every 100 molecules that enter the body, 10 molecules of metabolite get to the target site. But at higher doses, for every 100 molecules that enter, 50 molecules of carcinogenic metabolite reach the target site. At the lower dose (dose 1), the body somehow more efficiently rids itself of the dangerous metabolite than it does at the higher doses. This is not uncommon. If risk is a function of target site doses, then lower doses create less risk per unit of administered dose than do higher doses. If this is true, application of a linear model to the high dose data (based on administered dose) would overstate the degree of low dose risk.

Pharmacokinetic studies can yield information about the actual relationships between administered doses and target site doses over a range of doses; from this it may be possible to conduct a risk assessment based on the target site doses, and construct a more accurate profile of the dose–response relation.

Although these types of data are beginning to be incorporated into regulatory assessments, they are not always easy to collect and interpret. So, while we can expect wider use of ADME and mechanistic data in risk assessments in the future, for the present the regulatory principles mentioned earlier are the basis for most risk assessments issued by regulatory agencies.

Of course, not all scientists agree that these principles reflect current understanding. Some are more widely accepted than others – the linear, no-threshold model is particularly controversial, particularly for carcinogens that are not genotoxic – but regulatory agencies feel compelled to adopt such principles so that they can get on with the business of assessing risks, and ensure they are not likely to underestimate human risk. It is not in the least surprising that scientists can be found to cast doubt on the reliability of risks described by regulatory agencies, or to accuse those agencies of overstating the problem and creating unnecessary panic. And, of course, there will be others who attack on the ground that risks may have been understated and that agency actions will jeopardize health. The absence of a scientific consensus on critical aspects of the risk assessment process explains why the credibility of regulatory pronouncements about risk is always under attack. Regulatory risk assessors, indeed risk assessors everywhere, may also be faulted for failing to provide to the public a more thorough understanding of the uncertainties associated with their assessments. Pressures to avoid forthright discussions of uncertainties come from policy-makers who believe that the public does not take kindly to this topic, and will settle only for firm answers.

Threshold agent risks – the acceptable daily intake

The prevailing hypothesis that a threshold exists somewhere along the dose–response curve for all forms of toxicity except carcinogenicity, leads to an inquiry about its location for the human population that is to be protected. If a reasonably thorough set of toxicology data exists, and includes dose–response relations, it will be possible to identify the NOEL – the maximum dose (from the particular set of data revealing the most sensitive species and target for a chemical's toxicity) at which no adverse effects are observed (Chapter 9). But consider the discussion of the last chapter concerning the strong possibility that the threshold dose for human beings varies among them. Because this is the case the NOEL, particularly if it is based on data from studies involving small, homogeneous groups of animals, is probably not the threshold dose for a large and highly nonhomogeneous human population. It is common to estimate the threshold dose for the human population by dividing the NOEL by what are variously called 'uncertainty factors' or 'safety factors.' If, for example, the NOEL for a chemical derived from a chronic toxicity study is 100 mg/kg b.w./day, the threshold dose for a large human population is typically set at 1/100th of that value, or at 1 mg/kg b.w./day. The latter figure has historically been called the ADI – the acceptable daily intake.

While there are good reasons to believe that a safety factor of some sort is necessary to establish a human ADI, there are few solid scientific reasons for the selection of specific factors. In the early 1950s, two FDA scientists – Arnold Lehman and O. Garth Fitzhugh – were exploring methods for setting protective levels for human intake of food chemicals, based on toxicity data. These scientists reviewed the relatively scant literature available at that time on the relative sensitivities of humans and test animals, and the degree of variation of sensitivity within the human population. Based on a reading of the available literature, they concluded that a safety factor of 10 for each of these two variations – relative sensitivities of humans and rodents, and variation in sensitivities among humans – would be adequately protective. So they proposed that ADIs for chronic human exposure to chemicals, as might occur, for example, for food additives or pesticide residues in food, be set at 1/100th the chronic rodent NOEL.

The Lehman–Fitzhugh approach has been very widely used for setting limits on exposures to chemicals, not only in food, but in all other environmental media. There are several different safety factors used, the magnitudes of which depend on the nature and quality of the data available to establish a NOEL. The EPA has recently turned to using the term 'uncertainty factor' for those factors whose basis is a true scientific uncertainty, and distinguishes these from 'safety factors,' which reflect the injection of policy judgments that go beyond scientific uncertainties in the establishment of 'acceptable' intakes. EPA has dropped the term ADI, and instead calls the derived human threshold estimate an RfD – toxicity reference dose – removing the inference of 'acceptability,' which, they say, carries with it the connotation of a non-scientific, value judgement. In most cases ADIs (still used by FDA and the World Health Organization) and RfDs are identical.

What can be said about the protective value of the ADI? There is no way to be sure how close it is to the true threshold, by which is meant the threshold dose for the most sensitive members of the human population. If the generic safety factors are accurate, or are larger than they need to be, then the ADI should be adequately protective, or even more protective than it needs to be. A recent review of the available data on relative sensitivities of humans and animals, conducted by two scientists at EPA, suggests that these factors are adequately protective. But there is no completely rigorous way to pin down these values. Until the experimental scientists can offer a means to test relative human and animal sensitivities, and to test variability among humans, it will be necessary to continue to rely upon this relatively crude 'safety factor' approach.

Let us add, though, that there's no evidence that standards derived from the use of such safety or uncertainty factors are not protective. That is, there's no evidence that people whose intake of a chemical does not exceed its ADI have ever been harmed. We also need hasten to add, of course, that there are few good opportunities to determine whether harm has been created. We distinguish here, by the way, between the reliability of the method used to establish an ADI, and the reliability and thoroughness of the toxicity data upon which it is based. Obviously, if the toxicity data base is seriously deficient, or if important toxic effects have not been discovered, then the NOEL is inadequate and so is the ADI derived from it.

The ADI is not a sharp dividing line between 'safe' and 'unsafe' exposures. It appears likely that it will fall below the true but unknowable threshold dose for most people, but there's no way to guarantee that it covers absolutely everybody. The ADI is not an 'absolutely safe' intake level. Of course, we can not be sure that any exposure, or any activity, is 'absolutely safe.'

Notice that the ADI is not a direct expression of risk. Risk, recall, is a probability; the ADI is a 'very low risk' intake, or dose, with 'very low' undefined. If there were some means to estimate the fraction of the population having thresholds lower than the ADI (the potentially 'at risk' group), then a measure of the risk associated with the ADI could be obtained. No means exist to do this except for a few agents, particularly those such as the common air pollutants (ozone, carbon monoxide, sulfur and nitrogen oxides) that cause readily observable damage to human health.

Without knowledge of the risk associated with an ADI, and no good data or even hypothesis about the dose–risk relationship above the ADI and below the region of observable toxicity (i.e., the zone covered by the safety factor), then it is not possible to say much about the risk that exists when people are exposed above the ADI. Risk assessors can make only qualitative statements to the effect that risk increases at doses above the ADI, but in an unknown fashion; the greater the exceedance of the ADI, the greater the potential risk – the greater the fraction of the population whose thresholds are exceeded. Keep in mind that there are no doubt some ADIs that are excessively protective, such that doses well above them create no risk; it is just not easy to determine which ones those are.

Using ADIs

The final issue that needs to be mentioned here concerns the relationship between the ADI for a chemical and its associated level in an environmental medium. The ADI is a dose, typically expressed in mg/kg b.w./day. Consider mercury, a metal for which an RfD of 0.0003 mg/kg b.w./day has been established by EPA, based on certain forms of kidney toxicity observed in rats (these are not the only toxic effects of mercury, but they are the ones seen at the lowest doses). Suppose a limit on mercury levels in drinking water needs to be set. The goal is to ensure that the RfD is not exceeded. To do this, EPA first selects a hypothetical, average person, whose lifetime body weight averages 70 kg and who drinks the average two liters of water each day. If the RfD is 0.0003 mg/kg b.w./day, then the allowable daily mercury *intake* is:

$$0.0003 \times 70 = 0.021 \text{ mg/day}$$

That is, a 70 kg person could take in 0.021 mg of mercury each day, and thereby receive a dose of 0.0003 mg/kg b.w./day. If the 0.021 mg mercury were received entirely through drinking water, and two liters of water were consumed each day, then the 'safe' drinking water concentration (with rounding) is:

$$\frac{0.021 \text{ mg}}{2 \text{ l}} = 0.01 \text{ mg/l}$$

A drinking water concentration of mercury of 0.01 mg/l gives rise to an intake of just under 0.021 mg of mercury each day, which corresponds to a dose of 0.0003 mg/kg b.w./day.

This is so far simple, but there are complicating factors. First, drinking water will not be the only source of mercury exposure; some gets to people through certain foods, for example. So if the amount present in and taken in from other media is not considered, then the ADI could be exceeded if the drinking water level of 0.01 mg/l is accepted. The problem of multiple sources of exposure to the same chemical can be much more complex than is suggested in this example, but can not be ignored if the ADI/RfD is not to be exceeded.

Carcinogenic risks

Under the regulatory risk assessment model, the result of applying the linear, no-threshold hypothesis to carcinogenicity data is an estimate of the upper limit on what may be called the 'potency' of the carcinogen. By potency we refer to the upper limit on lifetime cancer risk associated with one unit of average daily lifetime dose, obtained by extrapolation as shown in Chapter 9, Figure 5. The carcinogenic potency is the slope of the straight line model A in Figure 5, at low doses. (Slope is the rise of that line – the increase in risk, shown in the vertical axis – for each unit rise in dose, shown on the horizontal axis.) The inherently more dangerous carcinogens exhibit steeper extrapolated slopes than do the less dangerous ones.

Under this model of cancer risk, the upper limit on lifetime risk can be estimated simply by multiplying the cancer potency by the number of dose units individuals are or could be exposed to each day. That is, if potency has units of 'upper limit on lifetime risk per unit of dose', and we multiply it by number of 'dose units', the result is 'upper limit on lifetime risk'.

EPA's potency estimate for chloroform, based on animal carcinogenicity data, is 0.006 in units of lifetime risk per one mg/kg b.w./day. (Human data regarding the carcinogenicity of chloroform are inconclusive.) One source of chloroform is drinking water, where it can form as a trace byproduct of the chlorination process used to disinfect water. Suppose the average chloroform concentration of a particular drinking water supply is 0.050 mg/l (EPA permits up to 0.100 mg/l, based on technical necessities). If people drink 2 liters of this water each day, a total of 0.10 mg chloroform will be consumed (0.05 × 2). If these people average 70 kg body weight, then the average daily chloroform dose will be:

$$0.10 \text{ mg/day} \div 70 \text{ kg b.w.} = 0.0014 \text{ mg/kg b.w./day}$$

If people receive this dose of chloroform each day for a full lifetime, then according to the EPA dose–response model, the upper limit on excess lifetime cancer risk is:

$$(0.006 \text{ risk per mg/kg b.w./day}) \times (0.0014 \text{ mg/kg b.w./ day})$$
$$= 0.000\,008$$

If this risk is accurate, it means that approximately 8 of every 1 000 000 people experiencing an average intake of chloroform through drinking water of 0.0014 mg/kg b.w., each day for a full lifetime, will develop cancer over a 70 year lifetime.

But remember, this risk (extra lifetime probability of cancer) is accurate only if all of the following hold:

(1) Chloroform is a human carcinogen.
(2) The animal carcinogenicity data provide an accurate picture of human response, in both the nature of the response (cancer) and its quantitative aspects (potency).
(3) The linear, no-threshold, dose–response model is accurate for very low exposures.
(4) The quantitative relation between administered dose and target site dose is the same at all doses.
(5) People actually achieve the estimated level of ingestion every day for a lifetime.

The estimated risk will be *greater* than the actual human risk if one or more of the following is correct:

(1) Chloroform does not increase cancer risk in humans under any conditions (in which case actual risk will be zero).
(2) The animal model is more sensitive to chloroform-induced carcinogenicity than are human beings.
(3) There is a threshold in the dose–response curve, or the curve drops toward zero risk more quickly than is suggested by the linear model. (curve B or C, from Figure 5, Chapter 9).
(4) The body is able to rid itself of carcinogenic chloroform metabolites more efficiently at very low doses than it can at very high doses.
(5) Exposure does not persist for a full lifetime or is otherwise less than that indicated.

The estimated risk will be *less* than the actual human risk if:

(1) Human beings are more sensitive to the carcinogenic effects of chloroform than are the experimental animals used for hazard and dose–response modelling.
(2) The actual dose–response curve falls toward zero risk more slowly than is indicated by the linear model (it is a so-called superlinear model).
(3) The body can not rid itself of carcinogenic metabolites of chloroform as efficiently at low doses as it can at high doses.
(4) Exposures are actually higher for some people because they are consistently exposed to water concentrations that exceed the average level.

Most scientists would hold that these unknowns and uncertainties in the regulatory risk assessment model would tend to favor risk overestimation rather than underestimation or accurate prediction. While this view seems correct, it must be admitted that there is no epidemiological method available to test the hypothesis of an extra lifetime cancer risk of about 8 per 1 000 000 from chloroform in drinking water. The same conclusion holds for most environmental carcinogens.

Risks have so far been presented in quantitative terms, with a discussion of some of the conditions that would have to be true if the risk were to be considered accurate. Additional commentary on the likelihood that these conditions are correct, and the likely effect on risk (to increase or decrease it) were any not to be, is a critical part of the risk characterization.

In most cases, the effect on estimated risks of alternative assumptions, and the probability that any such alternative is correct, can not be estimated quantitatively. So the risk characterization includes both quantitative expressions and descriptive commentary. Too often only the quantitative expressions are given much weight, both by decision-makers and the public. Numbers are easier to work with than is descriptive material, but this is no excuse for not trying to judge how close a particular estimated risk is likely to be to the true risk. The differences between risks estimated by the regulatory methods described and the true but unknowable risks will no doubt vary among chemicals. The data available for different chemicals are highly variable, and therefore so is the plausibility of the various science policy assumptions that are generically applied to all. The descriptive material accompanying the quantitative estimates should capture these important differences. Unfortunately risk assessors have not yet learned to articulate these matters very cogently.

Professor John Doull of the University of Kansas, a leading figure in many areas of toxicology, holds the view that toxicology, like medicine, is both a science and an art. Judgements about risk necessarily include factors that are very difficult to make explicit, but which are perceived to be true by experts who, depending upon their experience, have learned to weigh in some fashion large sets of data that can not easily be compared and evaluated in a completely objective way. Practitioners develop a 'feel' for the quality and meaning of toxicity data. This is no doubt important, thought it obviously sets the stage for major disagreements among experts with different experiences and perceptions. Nonetheless, many experts express opinions in this way, not untypical of medical judgements, about environmental risk matters that should have a more objective basis.

Professor Doull's characterization of toxicology is no doubt accurate; he is distinguishing, I believe, between the empirical aspects of risk assessment and what we have called the 'science policy' choices needed to complete a risk assessment. There appears to be no reason why the bases for the 'art', for the judgements risk assessors make, can not be set out in an explicit way, even though they do not merit the same scientific status as the facts that have been collected. Explicitness, even where science is weak, is one of the hallmarks of a modern risk assessment; nothing should be hidden behind the expert's cloak.

Actions based on hypothesized risks?

Many chemical risks such as those of chloroform in drinking water, are 'calculated,' not measured – that is, they are based not only on scientific data, but also on various sets of assumptions and extrapolation models that, while scientifically plausible (they fall within the bounds of acceptable biological theory), have not been subjected to empirical study and verification. Indeed, the results of most risk assessments – whether expressed as an estimate of extra cancer risk or an ADI – are scientific hypotheses that are not generally testable with any practicable epidemiological method. There is, for example, no practical means to test whether chloroform residues in chlorinated drinking water increase lifetime cancer risk in humans by 8 in 1 000 000, as hypothesized above. The tools of epidemiology are enormously strained, indeed, when called upon to detect the relatively low risks associated with most environmental chemicals. Without such a test, these risks remain unverified.

Regulatory officials nevertheless act on the basis of such hypothetical risks ('hypothetical' definitely does *not* mean 'imaginary;' it means that the risk estimates are based on certain scientific hypotheses and that they have not been empirically tested). Such actions are in part based on legal requirements and in part on prudence. The scientific information, assumptions, and extrapolation models upon which risk assessments are based are considered sufficiently revealing on the question of human risk to prompt risk control measures. To put off such actions until it is seen whether the hypothesized risks are real – to wait for a human 'body count' – is considered to be an unacceptable course.

The policy decision to act before science is certain does not, of course, dissolve the scientific uncertainties. Indeed, a strong argument can be made that an assessment of the type we have described should not pretend to represent normal science. Many of its outcomes are untestable with current methods; this alone might disqualify it as a true science, at least under philosopher Karl Popper's 'falsifiability' notion of the scientific method.

The counter argument to the above rests on the premise that risk assessment outcomes, if not currently testable, might be in the future. Many scientific hypotheses are not testable at the time they are proposed; this does not mean they are 'unscientific,' assuming they are based upon and are consistent with all available knowledge. Moreover, risk assessors continue to urge the development of the types of data that will improve the reliability and testability of their predictions; some of my own suggestions are offered in the final chapter. At the same time risk assessors, and the decision-makers who act upon their predictions, need to acknowledge that an element of policy invades their science, and that some of the bounds of normal science have been leaped over in an attempt to answer these exceedingly important questions.

As the scientific debate goes on, risk managers will no doubt continue to act upon incomplete information, because not to do so could be detrimental to the public health, and because the public will continue to call for answers about possible threats to its health and actions to protect it. Let us now explore more fully some of the risk management issues that arise when scientific uncertainty is pervasive.

11

Managing

Laws

Federal and state legislators in the United States have enacted laws that mandate certain types of controls on human exposures to just about every category of industrial chemical and environmental pollutant. Although these statutes vary regarding the degree of required risk reduction and the factors that need to be considered in the risk management decision, all seek protection of human health. Table 8 contains a list of some of the major laws, the categories of chemicals they cover, and a further notation regarding the factors that managers need to consider when making decisions.

The first thing worth noting about these laws is that they do not treat all sources of chemical exposures in the same way. Congress has directed the FDA, for instance, to consider only whether an added food ingredient is 'safe' when making decisions about allowing it to be used, whereas the same agency is allowed to balance the risks associated with a new drug against the health risks that might exist if the drug were not available for use in disease treatment. Under some laws EPA is allowed to consider both health risks and the availability of technology to control those risks, whereas others require the agency to consider health risks only when making decisions. OSHA, in setting workplace limits, is supposed to ensure worker safety and also consider the availability of technology to control exposures. The notations in Table 8 – risk only, technical feasibility, balancing – are provided as (somewhat oversimplified) guides to what decision-makers are required to take into account.

This is confusing. Why don't risk assessors simply decide what level

Table 8. *Some federal laws under which chemicals are regulated in the United States*

Law	Regulatory agency	Regulated products	Regulatory model*
Food, Drug and Cosmetics Act	FDA	Foods, drugs, and cosmetics, medical devices, veterinary drugs	Risk (food, cosmetics) Balancing (drugs, medical devices)
Federal Insecticide, Fungicide, and Rodenticide Act	EPA	Pesticides	Balancing
Federal Hazardous Substances Act	CPSC**	Household products	Risk
Occupational Safety and Health Act	OSHA	Workplace chemicals	Technical feasibility
Clean Air Act	EPA	Air pollutants	Risk (stationary sources). Technical feasibility (moving vehicles)
Clean Water Act	EPA	Water pollutants	Technical feasibility
Safe Drinking Water Act	EPA	Drinking water contaminants	Technical feasibility
Superfund Amendments and Reauthorization	EPA	Contaminants at waste sites	Risk/technical feasibility
Toxic Substances Control Act	EPA	Industrial chemicals not covered elsewhere	Balancing

* 'Risk' means the agency considers only risk information when reaching decisions. 'Balancing' means that both risks and benefits are considered. 'Technical feasibility' means that the law requires the agency not only to consider risks, but also the availability of technology to control risk. Some laws invoke more than one model.

** CPSC: Consumer Product Safety Commission.

Note: at the time of writing the Clean Air Act was reauthorized. The 1990 version is substantially different from the earlier law listed here.

of exposure is 'safe' for each chemical, and risk managers simply put into effect mechanisms to ensure that industry reaches the 'safe' level? Why should different sources of risk be treated differently? Why apply a 'no risk' standard to certain substances (e.g., those intentionally introduced into food, such as saccharin) and an apparently more lenient risk–benefit standard to unwanted contaminants of food such as PCBs and aflatoxins (which FDA applies under another section of the food law)? Why allow technological limitations to influence any decision about health? What is this risk–benefit 'balancing' nonsense? Aren't some of these statutes simply sophisticated mechanisms to allow polluters to expose people to risks?

These are pretty good questions, and they are not easy to answer, especially those concerning the different decision-making criteria associated with different laws.

Let us deal with this last issue first, and simply note that the various laws listed in Table 8 each have their own histories and were generally enacted quite independently. Their particular forms were fashioned out of a complex interaction of industry, consumer and environmental activist, and governmental constituencies that each brought its own agenda to the legislative process. It is not the purpose here to try to understand the differences that resulted, but rather to explore some of the effects of these differences on the problem of deciding what limits ought to be placed on human exposures to environmental chemicals. In the final chapter we shall suggest some possible changes in laws that might be desirable; for now we accept the existing legislative wisdom.

This chapter is by no means a comprehensive portrait of risk management issues. Many exceedingly complex technical and policy matters, to say nothing of the often volatile political factors, influence decision-making in particular cases. Emphasis here is on certain technical issues that arise in the use of the risk information that has been the subject of this book.

Safety

It should be clear by now that risk assessors do not know how to draw a sharp line between 'safe' and 'unsafe' exposures to any chemical. The very notion of 'safety' is scientifically wrongheaded, if by it is meant the absolute absence of risk. If 'safety' is defined in this way, it becomes in most cases impossible to know when it has been achieved, because to do so requires the proof that something – in this case, risk – does not exist.

That no risk exists can be proved under one and only one set of circumstances: when it is certain that exposure does not exist. How can the latter condition be ensured? The only real way is to guarantee that a chemical is not used for any purpose. We can know that exposure to cyclamate, a commonly used non-nutritive sweetener until 20 years ago when it was banned by FDA, does not exist (save, perhaps, in somebody's laboratory, where a few bottles might be sitting around, and in other countries, where its use is permitted), because food manufacturers are prohibited from adding it to foods and beverages. Cyclamate thus poses no risk to individuals in the United States – under present conditions we are absolutely protected from any risks this chemical may pose (a debate still continues about whether it is carcinogenic but that is irrelevant to this discussion).

But such 'absolutely safe' situations are not of much interest. While the use of some chemicals can be banned, it is not realistic to expect this approach to be applicable to all industrial chemicals or to the polluting by-products of industrial society. If the goal of absolute safety (zero risk) from these products is desired, then such a wholesale banning would be necessary. We do not appear ready to turn back the calendar 200 years.

To further this discussion let us divide environmental chemicals into three broad groups. First there is the enormous group of naturally-occurring chemicals that reach us primarily through food, but also through other media. Second are industrial chemical products that are produced for specific purposes. And third are the industrial pollutants – chemical by-products of fuel use, the chemical industry, and most other types of manufacturing.

Of these three groups the second is probably the easiest to control, in a technical sense. If society wishes to guarantee absolute protection from any of the forms of toxicity associated with these substances, then it will be necessary to prohibit their use altogether. If regulators were successful at banning all the products of industry, some of the pollutants arising as by-products of their manufacture (Group 3) would also disappear, but the large number of chemicals arising from other sources of pollution would still have to be dealt with. And, because it is not possible to ban food or the natural world, society would continue to live with the large number of chemicals from this source which, we have noted, pose risks of largely unknown magnitude because scientists have not paid much attention to any except those having serious, acute toxicities.

Because wholesale bans of these types will not occur, then another approach to achieving safety, at least for pollutants, might be suggested. Why not seek the goal of 'no detectable' chemicals in the media of human exposure? If automobiles omit various nitrogen oxides, simply ensure that emission rates are low enough so that these noxious chemicals cannot be found in air. If PCBs are migrating from a hazardous waste site, impose limits on that migration so that no detectable PCBs are found in the off-site environment. Control aflatoxin contamination of raw food commodities to ensure none can be found in finished foods.

This approach may sound pretty good, but it doesn't make much sense. The 'I can't find it, so it must be safe' approach to controlling environmental risks is flawed because it depends upon the operation of a relationship between technical capabilities to detect the presence of a chemical and the magnitude of the health risks it poses. There is no such relationship. Further elaboration of this issue is in order, and a specific example will be useful.

Diethylstilbestrol, mentioned in Chapter 7 as a synthetic estrogen that is also a human carcinogen, was used in the United States from the 1950s until 1979 as a growth promotant in sheep and cattle.[18] Small amounts of this drug, added to animal feed or implanted in the flesh of animals' ears, increase feed efficiency, and it was very widely used for this purpose.

Under the law FDA is charged with enforcing, carcinogenic substances such as DES can be used in food animal production, as long as 'no residue' of the drug is found in edible products, in this case beef.

This 'no residue' requirement of the federal food law seems to ensure safety. If there's no residue, then there's no exposure, and, it follows, no risk to anybody. Sounds perfect.

But what, exactly, does 'no residue' mean? If the applicable food law is examined more closely, we find it actually says 'no residue by a method of analysis' approved by the regulatory agency. This linkage of the phrase 'no residue' to a 'method of analysis' is important and suggests our legislators actually understood the necessity of establishing such a linkage (little chance this was actually the case).

Why such a linkage? Any analytical chemist will tell you that no method of analysis can ever reveal, under any circumstances, that a chemical is *not* present. If an analysis is performed on beef suspected of containing DES, and the chemical is not detected, the best the analyti-

---

[18] DES was used in poultry during the 1950s but FDA stopped this use in 1961.

cal chemist can conclude is that the compound was not present above the minimum concentration of the chemical the analytical method employed was capable of detecting. This level is called the 'detection limit' and varies from chemical to chemical, from one environmental medium to another, and from one method of analysis to another.

Up until the early 1970s analytical chemists could detect DES residues in beef tissue, specifically liver, at a level of 5–10 parts DES per one billion parts of beef (5–10 ppb). If DES were present above this level it could be detected with existing analytical procedures, but it could not be found if it were present at any concentration up to 5 ppb. Under conditions of cattle dosing approved by FDA, 'no residue' of the drug could be found in the late 1960s. The drug could be safely and legally used.

But, as they are always eager to do, research analytical chemists found ways to improve their procedures, and by the early 1970s they could detect DES residues at about 1 ppb and above. Guess what. DES could be found where none was detected with the earlier, less sensitive method, even though the drug was being administered to cattle at the same (approved) dosages.

This result was not surprising. Once a drug, or any chemical, enters an environmental medium, in this case animal tissue we use as food, some amount is going to be present. Although amounts may decline (as the chemical metabolizes or degrades, for example), it is not possible to conclude its concentration ever goes to zero. The best we can do is search for it with some method of analysis and if it is not found, conclude that it is 'not present above the detection limit' of whatever analytical method we use. If detection limits improve – become lower – it is expected that the chemical will be found where it could not be seen with the earlier, less sensitive detection procedures. The lower the detection limits, the greater will be the frequency of samples found to contain detectable concentrations.

What should FDA have done with the analytical data on DES? The law permitted use of the drug only if 'no residue' could be found. The agency acted in accordance with the law and initiated proceedings to ban the drug, an action not completed until 1979.

At the same time FDA was attempting to deal with DES residues, the agency also recognized the fundamental strangeness of the 'no residue' requirement. In effect, it said that if a carcinogenic animal drug could not be detected in food, the food was to be considered safe. This is odd, because it defines safety in terms of the capabilities of analytical chemistry. It is not only odd, it makes no sense whatsoever. Our ability to detect chemicals in the environment bears no relationship whatsoever to the degree of risk they pose.

In 1973 FDA proposed to make the law make sense. The agency recognized it had authority to specify the 'method of analysis' that must be used to ensure 'no residue'. Why not first specify, for a specific carcinogenic animal drug, the safe level for humans, and then require that the drug's manufacturer develop analytical methods that were demonstrated to be capable of detecting levels at least as low as the safe level? If 'no residue' of this drug were found in edible animal products using the method of analysis proved to be capable of detecting the maximum safe level, then the drug could be declared safe and legal. The problem was: how to define the 'safe level'?

At the time (early in the 1970s) the prevailing wisdom was captured in the convenient but somewhat misleading phrase 'There are no safe levels of exposure to carcinogens'. This phrase had been used by many experts on carcinogens in testimony offered to Congress on the occasion of its consideration of amendments to the basic food law, and in connection with other bills as well. Just what did this phrase mean?

Well, it was nothing more than a crude expression of the no-threshold hypothesis, as described in earlier chapters. Under this hypothesis, any exposure to a carcinogen increases the probability that cancer will occur. As we have emphasized several times, it does not mean that any exposure to a carcinogen will 'cause cancer'. If the exponents of the 'no safe level' view meant that an absolutely safe level of exposure could not be identified, then they were correct, assuming the no-threshold hypothesis is correct. (Even assuming the threshold hypothesis is correct, as we have earlier noted, does not establish that we can find the completely safe threshold dose for any agent.)

But if the definition of safety were to be converted to one that was not absolutist, then the 'no safe level' characterization no longer holds. FDA took the position that the 'safe level' for carcinogens such as DES was to be defined as that which produced no more than a specified, and very low level of excess risk. A risk assessment was to be carried out, based on animal carcinogenicity data obtained for the drug under consideration; the low, 'safe' risk level was to be specified; and the dose of the carcinogen corresponding to that 'safe' risk level was to be estimated from the dose–response curve. Figure 5, Chapter 9, shows graphically how this is done; the 'safe' risk level (on the vertical axis) is specified, and the corresponding 'safe' dose is estimated from the linear, no-threshold, dose–response model (curve A). FDA proposed a one in one million lifetime risk level as the maximum allowable, not the 10-fold higher risk of one in 100 000 shown in Figure 5 to be associated with dose A.

This approach was, in theory, more satisfactory than the absolutist approach, because it defined 'safety' not in terms of the scientifically meaningless and undefinable 'zero risk' standard (requiring banning, to ensure zero exposure, unacceptable as a general approach, as discussed earlier) but in terms that are scientifically meaningful because they do not require the impossible proof that something (risk) is absent. Safety, under this view, is a condition of very low risk.

Doesn't this pervert the meaning of 'safe?' Perhaps. It might in principle be desirable to cling to the popular definition of safety – no risk – as a goal, but we also need to face facts. First, that no activity or exposure, no matter how 'safe' it appears to be to the common sense, is demonstrably without risk. Whether we like it or not, we live with risk, it is unavoidable. There is nothing obviously wrong with seeking less and less risk (more and more safety), indeed many think there is a moral imperative to do so, but achieving zero risk is either not possible, or – and this is perhaps more relevant to the present discussion – is a condition that is not knowable. We shall note, but not further discuss, that seeking greater and greater reductions in risk in all circumstances is probably not wise public policy, given that we do not have infinite resources to spend on such reductions. Moreover, most risk reductions involve trading one risk for another; if this possibility is ignored, decision-making will be inadequate.

While it is clear that some people accept no definition of safety that is relative, it appears that most people feel safe when they are convinced that risks to their well-being are sufficiently low, even if not completely absent. (There are some dramatic and important qualifications on this conclusion, as we shall see in the later section concerning people's perception of risk. While for the most part people accept that the condition of safety is not equivalent to the condition of being completely risk-free, most people do not perceive risks as simply matters of probability, as do the experts. This intriguing and well-documented fact complicates greatly the public dialogue on matters of risk.)

While not everyone will be convinced that defining safety as a condition of very low risk is either wise or necessary, we shall proceed under the assumption that such a definition is the most technically sound one, and is the one that most people accept, either explicitly or implicitly. It is also the one that regulatory agencies have accepted, sometimes explicitly, sometimes not. FDA, as we showed, initiated this approach to carcinogens, in 1973. In fact this same approach had been widely accepted, at least implicitly, prior to FDA's actions. The methods long in use to establish ADIs for threshold agents cannot, as we have seen, guarantee absolute safety, even though they may appear to do so.

We were just about to deal with the issue of 'how low a risk is low enough' (how safe is safe?), before we embarked on the discussion of the definition of safety.

The FDA, borrowing the idea from the Mantel–Bryan paper mentioned in Chapter 9, declared that, for carcinogens such as DES, the condition of safety would be satisfied if the extra lifetime risk of cancer associated with consumption of residues in food did not exceed some very low level. In a regulatory proposal published in 1977 as a follow-up to the original 1973 proposal, the agency spelled out some of the reasons it selected a one in one million risk level as that level. In essence, the FDA held that if this risk were accurate, and if every one of the 240 million people in the United States were exposed daily, for their full lifetimes, to the residue concentration that created this risk, then the number of extra human cancer cases created over a 70-year average lifetime would be:

$$(240\,000\,000 \; people) \left(\frac{1}{1\,000\,000} \; \frac{extra \; lifetime}{risk \; per \; person} \right) = \frac{240 \; extra \; cancer}{cases \; during \; a \; lifetime}$$

or, an average of $240 \div 70 = 3$ to 4 extra cases per year (for an average lifespan of 70 years). The agency then noted that the model used to estimate risk (the linear, no-threshold model) would likely overstate the size of that risk (for the same reasons set forth in the last chapter); and, moreover, that it was very unlikely that anybody, let alone all 240 million citizens, would be exposed daily, for a full lifetime, to the maximum allowable food concentration of drug residue. So, the agency concluded, the actual number of extra annual cancer cases associated with a one in one million risk level, estimated using the regulatory risk model, would almost certainly be well below 3–4, by an undeterminable amount. Given that there are nearly 1 000 000 new cancer cases per year in the United States (exclusive of skin cancers related to sunlight), it appears that the one in one million risk level was an adequate definition of safety.

This was a policy, or risk management, choice on the part of FDA, pursued to seek a method for limiting exposure to carcinogens that would rest upon the degree of risk posed, not the irrelevant capabilities of analytical chemists. The same, one in one million *insignificant* risk level would apply to all carcinogens used as drugs for food-producing animals, and this would result in variable allowable food residue levels, depending on the relative potencies of the carcinogens (lower allowable levels for more potent carcinogens).

Expanding uses of risk assessment and the concept of insignificant risk

It became apparent to regulatory agencies during the 1970s that some means had to be developed to deal with carcinogens. Before that time carcinogens were either ignored, banned, or limited up to the amounts analytical chemists could detect. None of these was a very satisfactory approach to what was becoming obvious to all: more and more commercially important chemicals were being identified as human or, more typically, animal carcinogens; and chemists were finding these chemicals in more and more environmental media at lower and lower levels. Moreover, society was demanding, as evidenced by the spate of new laws passed in the United States in the 1968–1972 period, a clearer picture of the risks posed by these substances and greater control of those risks. At the same time FDA was promoting the use of risk assessment/insignificant risk determinations for the very limited class of carcinogenic animal drugs, the scientific literature began to see more papers on risk assessment methodologies. All of this led EPA to begin adopting risk assessment approaches for carcinogens, first in connection with pesticides, and eventually for all classes of regulated chemicals, including wastes found at Superfund and other such hazardous waste sites. OSHA, which at first rejected risk assessment as a basis for regulating workplace exposures to carcinogens, eventually adopted the technique. Even the Consumer Product Safety Commission got into the act, as it had to deal with carcinogens such as asbestos in certain hair dryers and formaldehyde in some home insulating materials. Many state regulatory authorities are also at it. To go into all of these regulatory uses of risk assessment would divert attention from the main theme: That there now exists a decision-making process for managing risks from carcinogens in the environment that includes the use of risk assessment and the further notion that human health can be adequately protected by ensuring that risks do not exceed certain low levels. Moreover, regulatory policy makers have emphasized that no single risk level, such as FDA's original one in one million level, satisfies the requirements of all the different laws that pertain to environmental carcinogens. Decisions about the appropriate risk goals for air pollutants, water pollutants, pesticide residues on food, food additives, occupational carcinogens, and so on, depend upon the requirements of applicable law, and policy makers have the responsibility to select risk reduction goals using the criteria set forth in these laws. Thus, as noted in the opening sections of the chapter, some statutes require zero risk for carcinogens (i.e., a complete ban on use, as required for those food chemicals covered by the Delaney Clause of the Federal Food, Drug and Cosmetic Act, but not required for other classes of food chemicals). Others require that risk not exceed some specified, insignificant level; and under other laws, agencies are permitted to consider the technical feasibility of various risk reduction techniques, and, under still others, some type of balancing of risks and benefits is permitted.

It is no wonder – and a source of confusion – that people can be exposed to different levels of risk from the same, regulated chemical depending upon whether they breathe it, drink it, consume it as part of their diets, or come into contact with it in their places of work or through the use of consumer products.

Moreover, under current laws, there is little opportunity for examining the totality of exposures from all media, which should be done because it provides a picture of total risk and allows identification of those sources that are the greatest contributors to that total risk. Risk decisions for chemicals in specific media are sometimes taken in isolation from decisions about the same chemicals in other media, and this practice is encouraged because of the diverse requirements of our laws and the artificial barriers they create. More on this topic is presented in the last chapter.

With some simplification, the current approach to regulating carcinogens goes something like this. Except for the zero-risk case of chemicals (intentional, directly-introduced food and color additives) covered by the Delaney Clause of the federal food laws, which requires banning, environmental carcinogens are controlled at different non-zero, risk levels. The FDA, which enforces the Delaney Clause, has taken the position that certain classes of food ingredients not covered by that clause, can be permitted as long as excess cancer risks do not exceed some very low (i.e., insignificant) level. Drug residues in animal products are, as we have seen, among the classes of food chemicals permitted under this regulatory model; FDA has also applied risk assessments to certain carcinogenic food contaminants (PCBs in fish, aflatoxins in peanut and other products) and concluded that risks greater than the one in one million level normally applied to added substances can be tolerated, because these contaminants are not completely avoidable. For these substances FDA applied a 'balancing' approach: the risk associated with the contaminants against the amount of contaminated food that would have to be removed from commerce at various tolerated levels of the contaminants. Aflatoxins and PCBs cannot simply be removed from food; foods containing excessive amounts of them have to be destroyed.

FDA also regulates food additives – substances, such as antioxidants, emulsifiers and non-nutritive sweeteners, that are intentionally and directly added to food to achieve some desired attribute. As noted the Delaney Clause prohibits the deliberate addition to food of any amount of a carcinogen. These additives, if they are threshold agents (not carcinogenic), can be allowed as long as the human intake does not exceed a well-documented ADI. Those who would seek approval for an additive need to supply FDA with all of the toxicity information needed to establish a reliable ADI, and all of the product use data that would permit the agency to assure itself that the ADI will not be exceeded when the additive is used.

Under EPA's interpretation of the Federal Fungicide, Rodenticide and Insecticide Act (FIFRA) the agency is allowed to balance the risks associated with use of a pesticide against the benefits that would be lost were the pesticide not available. For pesticides found to be carcinogenic, the agency has tended to use the one in one million lifetime risk standard, but departs from it to allow somewhat higher risks when benefits are judged high, and seeks somewhat lower risks when benefits are thought to be negligible. Interestingly the EPA tolerates higher risks for exposures to pesticides incurred by workers who manufacture, distribute, or apply pesticides than they do for the general population. ('Balancing' is, if anything, much cruder than risk assessment itself; methods for measuring pesticide benefits and balancing against health and environment risks are pretty much unexplored, yet FIFRA requires it.)

The toleration of higher risks for people who are exposed to chemicals on the job is not confined to pesticides. There is, in fact, a long tradition in toxicology to apply smaller safety factors (for threshold agents) when establishing protective exposure levels for workers than when establishing such levels for the general population. This makes sense, as a general matter, because workers are, on average, healthier than the general population, and the workforce does not include children, the infirm and the aged, and contains lesser numbers of individuals likely to be especially sensitive to chemical toxicity. And, at least in recent years, regulations have compelled the delivery to workers of extensive information about the chemicals they work with. In many cases their environments are monitored and, for workers handling especially hazardous chemicals, medical surveillance to detect early signs of a problem may be required. The Occupational Safety and Health Administration (OSHA), an arm of the Department of Labor, is the federal regulatory authority in these matters.

OSHA has in the past decade completed a number of rulemakings on occupational carcinogens, including arsenic, benzene, asbestos, ethylene oxide and acrylonitrile. The agency conducted risk assessments and concluded that occupational exposure standards – so-called Permissible Exposure Levels, PELs – were too high and had to be reduced.

The risks OSHA estimated were based on the assumption that a worker could be exposed to the chemical for a working lifetime of 40–45 years, and that exposure each day of that period would be the maximum level permitted, the PEL. Because these exposure conditions are unlikely to exist for any individual, actual job-related risks are almost certainly lower than the levels OSHA estimated, by unknown and varying degrees. Nevertheless, the excess cancer risks that OSHA found tolerable (in most cases because of the technical limitations on achieving lower exposure and risk levels) are greater than any that EPA or FDA has seen fit to tolerate for members of the general population. While some FDA and EPA risk decisions go as high as one in 10 000, most are at lower levels; OSHA decisions on occupational carcinogens have generally not forced lifetime cancer risks below about one in 1000.

It is hard to find compelling reasons to support the proposition that the workforce is less susceptible to cancer (as opposed to certain other forms of toxicity) than is the general population, so justification of the apparent double standard on these grounds is problematic. One of OSHA's considerations in reaching decisions about tolerable risk levels has been information on job-related risks of other types. The agency has cited Bureau of Labor Statistics data on job-related fatalities arising from accidents and other hazards unrelated to chemical carcinogenicity. OSHA found that lifetime risks of death associated with jobs most people perceive to be safe (office work for example) fall in the range of 1 per 1000, for a 40-year work period. The average lifetime risk of work-related death is 2.9 per 1000 in private sector establishments in the United States with more than 10 employees.

Work-related risks of death in construction, mining, lumbering, and agriculture are 3–10 times higher. This type of information was used by OSHA, together with arguments about technical feasibility, to support their decisions. Note also that OSHA requires some type of medical monitoring in cases where early detection of a developing cancer is possible.

EPA's Drinking Water Office sets limits on contaminants of, not surprisingly, drinking water, under the requirements of the Safe Drinking Water Act. This arm of the EPA establishes ADIs (RfDs) for chemicals that do not appear to be carcinogens, and then drinking water limits are set so that the ADI (actually a fraction of the ADI) is not exceeded, as in the example of mercury provided on page 196. But for carcinogens, say the drinking water regulators, the goal for exposure ought to be zero. Because this ideal cannot be achieved, limits for carcinogens – Maximum Contaminant Levels – are established at the lowest technically feasible level. This typically translates to a lifetime risk of one in 100 000 or lower, but for a few agents, risks associated with the Maximum Contaminant Level are greater.

EPA makes decisions about clean-up of certain hazardous waste sites under the so-called 'Superfund' law. Risk assessment outcomes are one guide to the decision process. The agency has declared that, for carcinogenic contaminants, clean-up must reach lifetime risks somewhere in the range of one in 10 000 to one in one million; most decisions seem to aim at risks of one in 100 000 or lower.

Curtis Travis of the Oak Ridge National Laboratory and several associates (including Richard Wilson and Edmund Crouch, who will be mentioned in a few pages in connection with our discussion of risk comparisons) reviewed 132 regulatory decisions on carcinogens. These authors generalize as follows:

(1) Every chemical exposure resulting in an estimated lifetime cancer risk of 4 in 1000 (one in 250) was subjected to some type of regulatory action.
(2) If the exposed population was very small (e.g., residents near an isolated manufacturing facility) lifetime risks below one in 10 000 never prompted regulation.
(3) If the exposed population was very large, typically the entire population, risks less than one-in-one million did not always prompt regulatory action, but risks greater than 3 per 10 000 always did.

Our review of some regulatory decisions, based on risk information, has been a relatively superficial one, and has avoided many complicated legal and policy issues, to say nothing of the political warfare that may accompany some decisions. The review, though selective in its

coverage, does reveal that regulators draw no single line of demar-
cation to separate 'safe' from 'unsafe' exposures to chemicals – i.e.,
that safety is a relative condition, defined according to the degree of
risk found tolerable under specific circumstances. We have also seen
that regulatory authorities tolerate a fairly wide range of risks, based
on varying legal requirements and historical precedents for specific
classes and sources of environmental chemicals; it is also apparent that
these differences in risk toleration are difficult for the public to
understand and, if truth be told, are not always easily understood by
close observers of the regulatory scene. There is much that might be
done to improve this risk management system, and we shall devote
some of Chapter 12 to some suggestions for such improvements.

Risk management as undertaken by federal regulatory agencies has
been the principal focus in this chapter, but similar decision-making is
fast becoming a component of corporate life. Many manufacturers and
users of chemicals have mounted programs to gain a better understanding of the risks their products pose – to workers, to consumers, to
individuals exposed to emissions to the environment, and to wastes –
and to undertake their own management actions, even when regulations have not yet demanded them. Careful and honest evaluations of
risk and the recognition that something must be done if risks are found
to be excessive are among the trademarks of environmentally enlightened corporations.

Experts versus the public

Risk managers are confronted with a host of fairly complex technical,
legal, and social issues when making decisions about whether to
restrict people's exposures to environmental chemicals, and about the
degree of that restriction. When all of this complex analysis is done,
however, the manager needs to be able to face the public and declare
that the final decision will ensure that its health will be protected. The
effective manager will have to explain why a particular risk level is
adequately protective and why it should be accepted by the public.
Whether it is a manufacturing plant manager explaining to members of
a surrounding community why they need not fear the emissions from
the facility, or a regulatory official explaining why certain pesticide
residues on tomatoes or a new food additive are safe to consume, the
issue is pretty much the same – the public needs to be assured that its

health is not jeopardized. And the public's understanding needs to be sufficiently sound so that it can enter the dialogue regarding the adequacy of decision-making. Public participation in the process would seem crucial to the success of any decision affecting it.

The risk manager needs both to be confident of the wisdom of the decision and to be able to articulate clearly why it was made. He or she needs to have a fairly good understanding of the risk assessment underlying the decision, most especially the uncertainties associated with it and how they were handled by the risk assessor. Confidence in the risk assessment, but not a foolish overconfidence, is essential. The manager must be assured the assessment represents the current state-of-the-art, and also needs to be able to explain why the current state-of-the-art, though imperfect, is the best that can be done. Some people may expect certainty from science, but most recognize that the quest for certainty is an illusion and will be put off by overconfident statements about what is known about risk, or its absence. Statements like 'we are sure this stuff is perfectly safe', even if uttered by highly regarded scientists (assuming such persons would ever make such a statement) inspire only mistrust.

The effective risk manager also needs to be able to explain why particular risk goals (or safety margins) were selected. Most people can be made to appreciate the impossibility of a risk-free environment, although no doubt there will always be some who refuse to accept this notion, at least for that part of the environment containing industrial products. At the same time people are not willing to have a risk imposed on them that they perceive as unacceptably high, and will challenge decisions that do not satisfy them in this regard. Here we enter an area of discourse that is problematical, to say the least.

Decision-makers have sometimes found presentations of *comparative risk information* a useful aid to the public discourse on risk acceptance. We referred in the last section, for example, to OSHA's use of statistics on the risks of job-related accidents to support decisions on risk reduction goals for workplace carcinogens. The agency noted that lifetime risks of death from injuries suffered in what most people perceive to be safe occupations do not go below about 1 per 1000. Data of these types were helpful in explaining why the agency settled on carcinogen risk levels in this range as sufficiently low to provide a safe work environment.

Professor Richard Wilson of Harvard University and an associate, Edmund Crouch, among others, have devoted considerable effort to collecting and analyzing risk information on activities commonly

Calculated Risks

Table 9. *Annual risks of death associated with some activities and exposures, as compiled by Edmund Crouch and Richard Wilson*

Activity/exposure	Annual risk (Deaths per 100 000 persons at risk)
Motorcycling	2000
All causes, all ages	1000
Smoking (all causes)	300
Smoking (cancer)	120
Fire fighting	80
Hang gliding	80
Coal mining	63
Farming	36
Motor vehicles	24
Rodeo performer	3
Fires	2.8
Chlorinated drinking water (chemical by-products)	0.8*
4 tbsp peanut butter/day (aflatoxin)	0.8**
3 oz charcoal broiled steak/day (PAHs, Chapter 7)	0.5
Floods	0.06
Lightning	0.05
Hit by meteorite	0.000 006

* Assumes water contains maximum level of by-product permitted by EPA; most water supplies contain less.
** Assumes aflatoxin present at maximum FDA-permitted level; most commercial brands contain much lower levels.
Source: Crouch and Wilson as cited by Slovic, P., 1986. Informing and educating the public about risk. *Risk Analysis* 6:403–15.
Note: Risks from activities are actuarial and much more certain than those associated with chemical exposures, which are estimated using regulatory models. Risks of cancer are assumed to equate with risks of death. Lifetime risk will be about 70 times higher if risks do not change substantially from year to year.

engaged in and exposures commonly incurred. This type of information can be effectively used both to educate the public about risk in general and to assist risk managers' efforts to explain specific risk decisions.

Some of the risk data assembled by Wilson and Crouch are presented in Tables 9 and 10. In Table 9 are lifetime risks for a number of activities and exposures that most people undertake or experience.

Table 10. *Risks estimated by Wilson to increase chance of death in any year by 0.000 001 (1 chance in 1 million)*

Activity	Type of Risk
Smoking 1.4 cigarettes	Cancer, heart disease
Spending 1 hour in a coal mine	Black lung disease
Living 2 days in New York or Boston	Air pollution
Traveling 300 miles by car	Accident
Traveling 10 miles by bicycle	Accident
Flying 1,000 miles by jet	Accident
Living 2 months in Denver on vacation from New York	Cancer caused by cosmic radiation
Living 2 months with cigarette smoker	Cancer, heart disease
One chest x-ray taken in a good hospital	Cancer caused by radiation
Eating 40 tbsp. of peanut butter	Liver cancer caused by aflatoxin*
Drinking 30 12-oz cans of diet soda	Cancer caused by saccharin
Living 150 years within 20 miles of a nuclear power plant	Cancer caused by radiation
Risk of accident by living within 5 miles of nuclear reactor for 50 years	Cancer caused by radiation

*Assumes aflatoxin at maximum FDA permitted level; most commercial brands contain much lower levels.

Source: Wilson, R. 1979. Analyzing the risks of daily life. *Technology Review* 81(4), 41–46.

Note that some of the risk information is actuarial (based on statistical data, typically collected and organized by insurance companies) and some of it has been derived from the type of risk assessment discussed in this book (chloroform in chlorinated drinking water, aflatoxin in peanut products). While the uncertainties associated with the figures in Table 9 are much greater for some risks than for others (not a trivial problem in presentation of risk data), such a presentation, it would seem, is very helpful to people who are trying to acquire some understanding of extremely low probability events, of the order of one in one million.

One of Wilson's more interesting presentations is that depicted in Table 10. Here exposures or activities associated with annual (not lifetime) risks of one in 1 000 000 have been described, another useful way to help people gain some sense of the 'reality' of very low probability events.

Comparative risk analysis is undoubtedly highly informative and can help risk managers to make decisions and then to explain them. But there is another issue here: it is quite clear from a good deal of research by social scientists that peoples' notions about risk are considerably more complex than those of the experts. *People do not perceive various threats to their health and well being simply as matters of probabilities.* Many attributes of a potential threat besides its probability of occurring influence people's judgments about whether they are willing to tolerate it, or, as Professor Peter Sandman of Rutgers University puts it, contribute to a determination of how much 'outrage' they feel. Woe to decision-makers who do not consider the implication of the work of investigators such as Sandman, and of Paul Slovic, Baruch Fischoff and their colleagues at Decision Research, in Eugene, Oregon. This is both fascinating and important.

Investigators such as Slovic, Fischoff and Sandman have studied how people perceive, or feel about, potential threats. A few of the attributes of a risk people consider, either explicitly or implicitly, in forming judgments, are the degree to which it is voluntarily assumed, the extent to which a personal benefit is perceived to exist as a result of incurring the risk, and the degree to which personal control over the risk is felt.

Generally, people are much more willing to tolerate risks that are voluntary than those they perceive as imposed on them. Professor Sandman illustrates this phenomenon with a simple example. The first-time skier is at fairly high risk of injury, but clearly accepts that risk. Now consider the person who is kidnapped, brought to a mountain-top, has skis strapped to his feet, and then is pushed off the top. On the way down the second person is nearly at the same risk of injury as the first, but she is certainly much more frightened than the first and also pretty angry about the situation – her perception of the risk that has been imposed upon her, and her feelings of threat and outrage, are qualitatively different from those of the first-time, voluntary skier. The objective facts about the degree of risk are not terribly important considerations for the second skier.

Although Sandman's example does not perfectly represent the whole of the issue of the voluntariness of the risk, it conveys a good sense of what makes it important. It is not surprising, then, that when they learn about benzene emissions to the air they breathe, people living near a petroleum refinery are not going to be easily satisfied simply by an explanation that the lifetime risks of cancer associated with those emissions are no greater than one in 100000, even though these estimates are probably pretty conservative, and even though they are more than 1000 times less than the risks of death from driving an automobile! Although people living in such circumstances are free to move away, most people would consider such exposures largely involuntary – they would be truly voluntary only if someone knew about them before moving to the affected area.

Degree of control is also important. The risks of riding in an airplane are perceived by many people to be much greater than the objective facts reveal. Part of the reason for this is that people tend to fear catastrophic events, such as a crash that may kill or injure many people at once, much more than they do events that take only one or a few lives at a time (such as accidents involving the very much riskier means of transportation, the automobile). But another part of the fear concerns the fact that people riding in airplanes feel they have absolutely no control over their fates. People driving automobiles feel safer than do passengers, for the same reason. Not rational? Perhaps, but nevertheless a common feature of human psychology.

Pesticides and food additives can provide many benefits, not only to food producers but also to consumers. But most people are not very aware of those benefits, or at least do not personalize them to a high degree, and this no doubt contributes to their sense of outrage when they hear about new health risks from these sources. Why should I take a risk when the only people deriving the benefits are the manufacturers? Some people take very high risks, whether on the job, as part of their recreational activities, or resulting from their personal habits (smoking, excessive alcohol consumption), because they feel they are getting something out of it for themselves; but they become upset when asked to tolerate very much smaller risks from activities or exposures that they feel are without significant personal benefits. Even if, in fact, they do derive some personal benefit, unless this is known to them, or is internalized in some way, people will not perceive it, and perception is what is important here.

Slovic, Sandman, and others have found numerous other attributes of a potential health threat to be important influences on people's perceptions. A threat that is of natural origin is more readily tolerated than a man-made one. Risks associated with familiar technologies are much less fearsome than those arising from new technologies (the products of biotechnology, for example). Some diseases or injury are perceived to be far more dreadful than others (cancer is certainly near the top), and risks that create such conditions are accordingly more dreaded. Even this brief sketch should prompt the reader to want to learn more about this extraordinarily interesting topic, so more references to some primary works in the field are provided in the bibliography.

These features of human psychology (I am not a social psychologist so I shall not attempt to deal with the topic of why people perceive threats as they do) are important factors in the public discourse on risks from chemicals in the environment. It is not difficult to discern that risks from synthetic chemicals in the environment, whether they be contaminants or useful industrial products, tend to be among those for which people have the least tolerance. Exceptions might be products that have been around a long time and that have come to be seen as important in people's lives. The most pronounced expressions of outrage brought about by revelations during the 1970s that saccharin, the artificial sweetener used since the turn of the century, was an animal carcinogen, came from those who did *not* want FDA to ban it (indeed, Congress passed a special law to keep it on the market!). But saccharin and products like it are exceptions to the general rule. So governmental and corporate officials who have to defend their risk reduction decisions regarding environmental chemicals have a doubly tough task. They not only have to be able to explain why particular risk levels are adequately protective – i.e., that absolute safety is not achievable and that it need not be achieved to protect the public health – but they also need to be able to deal with people's inherently low toleration for these types of risk. This is really hard, perhaps even impossible to accomplish with anything close to perfection, but it must be attempted. How do we get people to worry about and act upon the major risks to their lives, especially those that they do not perceive as particularly threatening, and to stop worrying excessively about minor risks? This is a tall order, but a very important one, and the type of understanding created by the research of Slovic, Sandman and others is immensely helpful to this effort. It is also helpful in preparing the way for wider public discourse on these topics, so that people can participate actively in decision-making. To get further into this issue – into the area of risk communication – moves us too far from the central purposes of this book. So we will stop here with a quote from William Ruckelshaus, twice Administrator of the EPA:

In a society in which democratic principles dominate, the perceptions of the public must be weighed. Instead of objective and subjective risks, the experts sometimes refer to "real" and "imaginary" risk. There is a certain arrogance in this – an elitism that has ill served us in the past. Rather than decry the ignorance of the public and seek to ignore their concerns, our governmental processes must accommodate the will of the people and recognize its occasional wisdom. As Thomas Jefferson observed, 'If we think they [the people] are not enlightened enough to exercise their control with a wholesome discretion, the remedy is not to take it from them, but to inform their discretion.'

12

Improvements and New Challenges

Scientists cannot respond fully to the social demand for knowledge about threats to our health from exposures to environmental chemicals. People obviously want to know whether their health is jeopardized by the tens of thousands of chemicals, particularly those of industrial origin, that are the ingredients of twentieth century life. This demand for knowledge is in part reflected in the many laws that require manufacturers, users, and disposers of chemicals to provide information to state and federal regulatory authorities that can be used to assess the health risks these chemicals may pose. Increasingly many corporations are finding it useful to undertake such evaluations on their own. Results of risk assessments are used to justify and guide risk reduction measures. All of this scientific analysis and risk management goes on under the watchful gaze of a number of consumer and environmental activist groups. It is a vast enterprise, filled with some good and some not so good science, generally uncertain technical analyses, major scientific and policy disputes, and seemingly endless litigation. Public discourse is sometimes reasoned, but often rancorous. The credibility of both industrial and governmental authorities, particularly those at the federal level, is increasingly doubted by the public. Individual states and local communities have in the past several years become much more active on these issues, not simply calling upon federal authorities to do more, but to do it themselves. The result has been a proliferation of regulatory and other risk management initiatives that American industry generally finds troublesome.

Perhaps the current state of affairs is healthy, and simply reflects the traditions of vigorous debate and demand for accountability that are the hallmarks of American life. But there also might be at work here some serious defects in the fundamental structures upon which the current system of chemical risk management is built which, if corrected, could help to move the debate to a higher level; to restore the credibility of all sides in that debate; to ensure that what government and industry are held accountable for is sensible and consistent with scientific understanding; and, perhaps, even to improve the public health. This chapter notes a few trends that the author believes need to be encouraged if the current situation is to be improved. But, unless the reader leave the book with the feeling that perfection is just around the corner, I also present a number of emerging issues that, if anything, are likely to create challenges even more difficult than those we now face.

Risk assessment

We have seen throughout this book that risk assessment is plagued by substantial gaps in knowledge. Although additional toxicity testing of chemicals is of undoubted importance, that testing is not by itself going to help us gain increased understanding of the two major uncertainties in risk assessment: extrapolation from animals to humans and from high to low dose. The only way to get at these matters – to reduce the extent to which we rely upon models for extrapolation and to increase the extent to which we rely upon data – is to gain increased understanding of the underlying mechanisms of chemically-induced toxicity, and to find ways to study the effects of low level exposures to chemicals directly in human populations, to improve the craft of epidemiological science. (Note that even if improved, epidemiology will be useful only for chemicals already 'out there'; experimental studies will still have to be used for new chemicals.)

Toxicologists, as we have seen, are turning more frequently to the tools of the fundamental sciences – biochemistry and molecular biology in particular – to learn in minutest detail how chemicals reach, interact with, and damage cellular components and how those forms of damage relate to the ultimate manifestations of toxic injury or disease. They are using cells isolated from human beings to study these processes, and are able to compare what happens in human cells with what goes on in animal cells exposed to the same chemical. Toxicologists are learning more about the dose–response relations that really matter by studying the relationships between the dose of the chemical that enters the body and the dose at the ultimate (typically cellular) target site for toxicity. They have interested molecular biologists in examining how specific types of chemically-induced cellular injury or changes evolve into the ultimate manifestations of disease. This is immensely exciting and important work.

Epidemiologists are seeking new ways to improve the sensitivity of their study methods. They are also finding ways to correct one of the major problems in current methods: learning about the extent of chemical exposure the individuals under study have incurred. Information about exposures may, for example, be gleaned by looking at certain components of the blood that have been chemically altered because of exposure. The blood of workers exposed to the widely used medical sterilant ethylene oxide, for example, contains some hemoglobin molecules that have reacted with the chemical. Chemical analysis of blood samples reveals the extent of this reaction and thus yields a measure of human exposure that is direct and potentially more telling than, for example, measurements of ethylene oxide levels in the air these workers breathe.

Results from these sorts of molecular toxicology and epidemiology studies have not yet paid off in a big way, but there is every sign that their contributions to our understanding of chemically-induced toxicity will be immense.

That is, if such work finds the necessary funding. In our search for understanding of chemical risks, there is a tension between the demand for resources for toxicity testing of more and more chemicals, and the need for resources to improve our understanding of what those test results really mean.

Those who believe that testing of chemicals should receive the higher priority argue that results from such tests are sufficient for decision-making about needed controls on exposure. We have a reasonably good system for risk assessment and all we really need to do is acquire more toxicity data to 'plug into' that system. Moreover, there are many new chemicals introduced into commerce every year, and many chemicals that have been in use for a long time, that have not been adequately tested. We cannot afford to go too slow on these matters, and that would be a consequence of our devoting too many resources to objectives other than the collection of toxicity information.

Those who favor greater attention and funding to those areas of research devoted to learning more about what toxicity test results really tell us about risks to human health argue that we could now be making serious mistakes in our assessments. Regulators, we have seen, fill the knowledge and data gaps in risk assessment by adopting, as a matter of science policy, certain assumptions (about dose–response, interspecies extrapolations, and so on) that are applied, generally, to all chemicals. What if these assumptions are wrong, if not for all chemicals then at least for many of them? Regulatory risk assessments might for some chemicals be artificially creating risks where none exists and for others they might be seriously understating risks. To base risk assessment and risk management decisions upon such uncertain scientific knowledge is bad public policy. In fact, the debate we now see among scientists, and among regulatory authorities, industry officials, and the public on matters of chemical risk exists largely because our knowledge is so insecure.

Both of these postures are meritorious, and the fact is that both research and testing need to be encouraged. We can surely go on making laws and regulations without a strong scientific foundation, but in the long term this is bound to cost us dearly, both in terms of dollars and human health. Our legislators and the public need to appreciate how far the scientific community is from being able to respond to their demand for knowledge about chemical risks.

Some scientists and public health experts will respond negatively to what I have just said. They will argue that we are already spending too much money and time on toxicity testing and research, because it is clear, they say, that environmental chemicals are not very important contributors to human disease. We saw in Chapter 7, for example, that the best evidence now available about the origins of cancer points to only a small role for industrial chemical products and pollutants, and to major roles for so-called 'lifestyle factors,' such as smoking, alcohol use, and dietary choices. It is wasteful to devote so many resources to the regulation of environmental chemicals, and to create so much public hysteria while doing so, because the public health benefits are minimal. Let's get on, they say, with the important tasks of educating people so they will change their behavior.

That changes in lifestyle factors are important in cancer prevention is indisputable, and efforts to get people to stop smoking and to change their dietary habits should be among the highest of our public health objectives. But there is too much not yet known about the role of environmental chemicals in human health; moreover, the public is demanding that we continue to seek that knowledge.

It is, first of all, obvious that cancer is not our only public health problem! Because of the application of the 'no-threshold' hypothesis to chemical carcinogens, this form of toxicity almost always overrides other forms in regulatory decision-making. This hypothesis, which is not without foundation for many carcinogens, confers greater risks at low, environmental exposures on carcinogenic responses than on those forms of toxicity – all the others – that appear to require exceedance of a threshold.

I suggest that the 'threshold/no-threshold' dichotomy applied in the regulatory risk assessment model probably greatly over-simplifies reality, and tends to create an overly narrow view of the possible implications for human health of what is known about the full range of toxic responses produced by environmental chemicals. Let's expand on this.

There are hundreds of unexplained human diseases that are almost certainly non-infectious, affecting all the organs and systems of the body, including our mental processes and behavior. People are also daily exposed to tens of thousands of environmental chemicals (I include natural chemicals here, a matter I shall return to in a few paragraphs); the small fraction of these that has been subjected to toxicity testing contains many substances that are capable, under some conditions of dosing, of causing toxicity to all of these systems and organs of the body. What we do not know, except for a few of these substances, are the relationships between the types of toxicity produced by chemicals, and the actual human diseases doctors are treating. The regulatory system does not inquire much into those possible relations – it simply assumes that if human exposure to a chemical is limited to a small fraction of that which is known to produce toxicity in animals, then the occurrence of adverse effects in humans will be prevented. This is no doubt important, especially when our knowledge base is so weak, but it does not go very far toward telling us how important environmental chemicals are in causing or contributing to the many human diseases of unknown origin.

Before telling scientific inquiries can be made into this question, it will be necessary to expand our concept of the phrase 'environmental chemicals.' Most people think the term refers to the products and by-products of chemical technology. This is a much too restricted view for the public health scientist, and impedes the quest for knowledge.

As we have noted several times in this book, the largest number of environmental chemicals people are exposed to on a regular basis are those present as natural components of the foods, beverages, herbs, and spices that make up their diets. In addition to the nutrients, there are hundreds of natural dietary chemicals that impart flavor, odor, and color. Thousands more are present simply because they play some role in the life of the plants, animals or microorganisms that comprise our diets. Only a relatively small fraction of these chemicals has been identified by chemists, but those that have been characterized display a very wide range of structural types, more varied and complex than most industrial chemicals.

Human beings have eliminated over the course of time many possible food sources because they were found to contain ingredients that were too poisonous. But we are referring here primarily to acute, readily observable poisonings. There has been no systematic investigation of the toxicities of natural food constituents that would be produced from chronic exposures. Of the few natural chemicals that have been investigated in this fashion, a large fraction has proved capable of producing the same range and forms of toxicity that has been found for synthetic chemicals. This is no surprise. Indeed, it is safe to conjecture that if toxicologists were to subject the natural chemical constituents of the human diet to the same type of chronic toxicity testing now used for industrial chemicals, a large fraction will be found to produce serious forms of toxicity, including cancer, birth defects, nervous system disorders, and so forth. In a lecture on risk assessment I presented at the New York Academy of Sciences in 1980, I concluded with the following:

It seems clear to me that as we begin to apply our current methods for detecting risks to some of the myriad natural components of food, it is highly probable that we shall uncover an uncomfortably large number of chronically hazardous substances. Although this should not shock anyone who understands the chemical nature of food, it will surely come as a surprise to the public. I suspect that at least two monumental dilemmas will surface:

(1) The public's skepticism about the adequacy of current methods to measure risk will rise, perhaps to unmanageable proportions.
(2) If we have high confidence in the methods we now use to measure risks from synthetic chemicals, and if we act on the basis of such risks, how can we justify not applying equally vigorous actions to the management of risks from the natural components of food?

I hope my conjecture about uncovering numerous and hitherto unrecognized risks among the natural components of food is incorrect so that we shall not have to face these dilemmas. But I fear I am right.

Professor Bruce Ames of the University of California at Berkeley, who we mentioned in Chapter 8 in connection with our discussion of the Ames Test, has devoted a great effort over the past several years to promoting scientific and public interest in this topic.

Ames points out that plants, including those we use for food, possess biochemical mechanisms that permit them to produce their own, natural pesticides, to ward off attack by insects and fungi. Ames has surveyed the vast literature on this subject and has identified a number of animal carcinogens among 'nature's pesticides'. He also finds that the concentrations of these natural pesticides are usually measured in the hundreds or even thousands of parts-per-million, whereas most man-made pesticides are present in the very low ppm or ppb range. Ames says, 'We estimate that Americans eat about 1500 mg/day of natural pesticides, 10 000 times more than man-made pesticides.' He also expects that, as with synthetic chemicals, if these products were to be tested for carcinogenicity at the MTD, as it is currently estimated, about half would turn out to be carcinogenic. Recall, however, that Ames is also of the view that testing at the MTD often creates extensive toxicity and cell proliferation, which puts cells at increased risk of progressing to a neoplastic state, such that bioassay results are not to be trusted as applicable to low (non-toxic) doses.

Ames' work has not yet caught on in a major way, indeed many actively resist his effort because they believe it will avert attention from industrial pollutants. But it needs to attract more attention from public health authorities, although it certainly need not mean an end to our concerns for industrial products.[19]

If we expand our notions of environmental chemicals to include natural components of the diet, we might then begin to see why it is of great importance to continue to study their health effects. Diet, as we have seen in Chapter 7, importantly contributes to human cancer rates. We do not understand a great deal about why this is, but it is surely far more complex than the rate and form of our fat and fiber intakes. Perhaps industrial products are not the major human carcinogens, but natural products may well be. To make matters more complicated, it is also apparent that many natural dietary constituents *protect* against certain cancers. Indeed, some may increase the risks of certain cancers and reduce that of others! It is not unexpected that some industrial chemicals will be shown to have these same properties.

We know very little about the relative roles of natural and industrial chemicals in the many other non-infectious human diseases; we may

[19] Professor Ames seems fully convinced that carcinogens of industrial origin, indeed industrial products generally, are of minor public health importance. Although his documentation and arguments in favor of an important role for natural dietary carcinogens are formidable, they seem not fully adequate, for reasons I discuss in the following paragraphs, to dismiss concerns about a significant role for industrial products and by-products in human diseases.

find that some industrial chemicals are important contributors to some diseases and natural chemicals to others. But it seems evident that the thousands of environmental chemicals that enter our bodies everyday are interacting in many different ways with cellular constituents and other body chemicals, and that some of these interactions initiate or promote processes that lead to various disease states. We are a very long way from understanding much of this, but we need to try.

Having said this, I suggest that the types of toxicity testing now used for environmental chemicals probably do not yield the best information for creating understanding of the public health risks they pose. They can no doubt give us important clues, but full understanding, it would seem, will require a deep appreciation of underlying mechanisms of toxic action, and of the biological relationships between those mechanisms and the ultimate manifestations of disease. Moreover, and this is one of the greatest conundrums in all of toxicology, we need to understand not only the effects of exposure to single chemicals, but to the combined effects of all the environmental chemicals to which we are exposed, again emphasizing the need for the broadest possible view of the chemical environment. There are a few notable examples of toxicological interactions among chemicals, some in which combined effects are greater than what would be expected if the effects were simply added (this is called *synergism*, and is well-represented by the case of asbestos exposure in combination with smoking), and some in which combined effects are *antagonistic*, that is, less than expected. No means exist to study toxicological interactions in a significant way, to cover more than a few chemicals in combination. As seen from the earlier review of mechanisms of carcinogenesis, nutrient interactions with toxicants are terribly important.

We clearly need to refine greatly our scientific tools, both epidemiological and experimental, for uncovering the role of our natural and synthetic chemical environment in human disease. And, so that we are not accused of putting only a negative slant on this matter, we need to point out that there is substantial evidence that some natural chemicals, and no doubt some synthetic chemicals as well (in addition to those we use as medicines) have a role in disease prevention. This means that our investigatory tools need to be able not only to sort the risky from the non-risky, but also to identify how certain chemicals actually reduce risks.

This is a tall order indeed and, in fact, it is probably not useful for planning research strategies to think in such broad terms as 'the role of environmental chemicals in human disease causation and prevention'.

I am attempting here only to create a shift in how we think about the chemical environment, and to broaden the horizons of the toxicological sciences.

Managing

The possibility that what most people perceive as environmental chemicals – industrial products and by-products – are not large threats to human health does not necessarily lead us to the view that regulatory agencies are currently excessively regulating them. Several factors from our earlier discussions need to be recalled.

First, although there is substantial evidence that many cancers are primarily associated with so-called 'lifestyle factors', the natural environment including viruses, and even genetics, we certainly cannot rule out some role for industrial chemicals. Moreover, chemicals can cause many other types of adverse health effects about which we do not have significant explanations of any type. Until we have evidence to rule out a significant role for industrial pollutants in human reproductive disorders, diseases of the nervous system, and so on, it would seem imprudent to cease or reduce gathering toxicity data on them and establishing limits on human exposures.

A second factor that obviously can not be ignored are the requirements of our many environmental laws. Although some scientists may argue that these laws place emphasis on the wrong risks, they nevertheless exist and must be enforced. Whether or not they have a strong scientific basis, they certainly express the will of many people that industrial products and the environmental contaminants resulting from their uses need special attention and controls. 'I don't care if a few ppb of trichloroethylene is not likely to create a significant health risk – I don't want any in my water!' This is a not uncommon sentiment and can not be ignored or attributed only to ignorance. Our laws are certainly not based solely on such views, but all contain an element of this type of thinking.

At the same time scientists have the responsibility to remind policy-makers and the public that the universe of chemicals does not divide neatly between the 'toxic' and the 'non-toxic'. Such false notions foster the view that all our problems could be solved if we simply identified and then rid ourselves of 'toxic' chemicals, the obvious first candidates being those of industrial origin. This simple and simple-minded view is much favored by those who can deal with issues only if they can be reduced to black and white terms, and elements of it can be found in certain laws calling for the development of lists of toxic chemicals. If this book teaches anything, it is that 'toxicity' is not equivalent to 'risk', and that degree of risk should guide public health decisions. Because risk assessment is imperfect does not mean we should turn to far less perfect guides to public health protection. So if we regulate based solely on some criterion of 'toxicity,' let us not pretend we know anything about the public health benefits we might have achieved.

Regulatory agencies probably can not shift their present course in a major way; indeed, there are significant arguments to be made that the agencies are not moving aggressively enough to enforce existing laws. Moreover, even those scientists who believe that the available evidence does not point to a significant role for industrial chemicals in human disease would support the view that we need to limit human exposure to these chemicals in some way – uncontrolled exposures to substances that can cause toxicity under some conditions is hardly justifiable. The question is: how much control is really necessary?

Our laws generally seek a very high degree of safety, that is, a very low degree of risk. As long as this is the case we can not avoid engaging in risk assessment, because directly measuring such low risks is beyond our current capabilities for most environmental agents. Risk assessments contain many uncertainties, most of which can not be rigorously quantified. Currently decision-making tends to ignore many of these uncertainties. It would seem that a significant improvement in our decision-making on environmental matters could be had if risk assessors did a better job of expressing uncertainties, and risk managers found some systematic ways to take them into account. Some models from the domains of decision-analysis might prove powerfully useful to regulatory and corporate risk managers. Finding ways to deal with a lack of scientific consensus, and to accommodate a range of scientific views, and yet to make practical use of the important information provided by a thorough risk assessment, is a task risk managers need to confront more aggressively.

Managers also need to improve their communications with the public on matters of risk. For reasons we reviewed in the preceding chapter, public understanding of risk information is generally poor and the public's perceptions do not match the judgments of experts. The effective manager needs to appreciate these matters and keep them firmly in mind when planning to explain decisions to the public.

And how much risk is too much? We have reviewed a few regulatory decisions about 'significant' or 'insignificant' risks. But it is hard to find strong analytical support for these conclusions about allowed and disallowed risks. Fundamental questions have not been significantly examined. What are our long-term societal objectives for risk reduction? Do we want to reduce risks endlessly, as long as we have the technical ability to do so? Do we give less attention to demonstrably high risks because it is not obvious how to reduce them, or because it is politically difficult, and instead devote resources to small, technically controllable risks, even though those resources might sometimes be disproportionately large compared to the actual public health benefit?

Getting at such questions in order to establish public health priorities and risk reduction goals in a rational way is greatly hampered by the fragmented regulatory structure that has been created to enforce our many environmental laws. These laws deal with different environmental sources of chemicals in different ways. The regulatory approaches to risk reduction thus vary depending upon whether a chemical is a food additive, a food contaminant, a pesticide, a drinking water contaminant, an air pollutant, or is, in fact, several of these. While each of these laws and sets of regulatory precedents have their own histories, and probably make sense within the context of those histories, the total picture concerning environmental chemicals and their regulation is almost impossible for anyone but the dedicated scholar to discern. Given where we have been, it is difficult to conceive how a simpler and more uniform risk identification and control system, generally applicable to all environmental chemicals, might be created, but at the same time it is hard to see why our policy-makers in government should not begin thinking about this issue. If one ponders the matter a bit, it seems astounding that we should tolerate such a complex, fragmented, and non-uniform system for controlling environmental chemicals.

As a final point we need to focus attention on a critically important risk issue that has been entirely neglected in this book, and that is only beginning to draw the attention it deserves. Our concern in this book has been focused on the effects on human health of exposures to environmental chemicals. We have not discussed how these chemicals may damage non-human life forms and even the inanimate environment (e.g., the ozone layer). This is an immense topic about which information is limited, but which could, in the long term, be more important in several respects than the topic of this book. An associate of mine has remarked that, somehow, the 'E' has been taken out of EPA, suggesting that the agency has devoted much more attention to human health protection than to environmental protection. The lack of attention to the non-human environment in this book is by no means an indication that it is unimportant.

A typical concluding paragraph in a book of this sort would carry an appeal for 'more research.' I hope by now such an appeal is unnecessary, because the need is simply obvious. It is hard to identify any area of science that figures as importantly in the formation of major public policies that has the degree of uncertainty we find in chemical risk assessment. Although we no doubt need to act prudently as long as the uncertainties are large (and we shall no doubt continue to argue about just how prudent we should be), this is not a satisfactory state-of-affairs, because we shall never know how much we are being deceived. Risk assessment is too important to be based heavily upon untested scientific assumptions that can be imposed to twist results one direction or the other to achieve desired policy objectives. The only way to test assumptions is to turn to the methods of scientific research.

But I do not want to end this book with a discussion of research issues. I want to bring up the topic of education. My sense is that matters of the environment, and particularly of risks created by it, receive relatively little attention in the curricula of our schools and colleges. Every poll I have seen shows that our population, and students in particular, are intensely interested in environmental issues. Everyone, it is safe to say, is greatly interested in the risks that threaten their health and well-being. High public interest, enormous social importance, significant scientific and legal content: what better criteria for establishing risk analysis as a significant item for inclusion in our educational curricula? I am not expert on mechanisms for accomplishing such an objective, but I suggest the public debate on these matters will not improve much until our population is better educated on the fundamental issues. A more aggressive curriculum might also serve to inspire some of our best minds to enter the environmentally-related professions, where they are sorely needed.

In the opening pages of this book I promised to clarify some of the major issues in risk assessment. I did not, however, promise to reveal the 'truth' about the risks associated with environmental chemicals. This would have been foolhardy, because we don't know the truth. We have evidence pointing in certain directions, we have methodologies for developing some information about risks, and we have significant precedents for basing decisions on that evidence and risk information. My goal has been merely to create some appreciation of the scientific basis for our present concerns, its strengths and weaknesses, and of how and why sometimes seemingly unfathomable public policies flow from those concerns.

Sources and Recommended Readings

Material for this book has been drawn from several basic reference works in toxicology and risk analysis, as well as from a number of primary scientific publications. The reference works are listed below, with commentary regarding their general content and utility. Many are rather expensive volumes but are readily available in university and some large public libraries. Also listed are certain of the primary sources that are of significant historical or technical interest, and that should be reasonably understandable to readers without much background in the sciences. These listings and commentary are primarily for the benefit of readers interested in further study of one or more of the topics covered in the book. Also noted are several public and private organizations that can be consulted for publications on chemicals and their hazards and risks.

General reference works and other sources of information

Casarett and Doull's Toxicology: The Basic Science of Poisons, 3rd Edition, is a multiauthor volume edited by Curtis D. Klaassen, Mary O. Amdur, and John Doull (Macmillan Publishing Company, New York, 1986) and is probably the single most widely-consulted general textbook in the field of toxicology. A comprehensive, multichapter unit on 'General Principles of Toxicology' is followed by detailed chapters on each of the major toxicity targets, classes of toxic agents and environmental pollutants, and on the applications of toxicology to regulation, occupational health, and medicine. Many of the chapters, particularly those concerning targets, are heavy going and require substantial training in biology, but there is an enormous amount of useful information for less-trained individuals; a good index makes finding that information relatively easy. Less difficult, but also substantially less comprehensive, is the single-author work, *Basic Toxicology*, by Frank C. Lu (Hemisphere Publishing Corporation, Washington, D.C., 1985). This volume is a particularly useful introduction to the principles and methods of toxicity testing. Kenneth Rothman's *Modern Epidemiology* (Little Brown & Co, Boston, 1986) is a clear, concise survey, particularly good on the problem of causation. Dennis J. Paustenbach has edited *The Risk Assessment of Environmental Hazards: A Textbook of Case Studies* (John Wiley & Sons, New York, 1989). Most of the material relating to basic principles of toxicology and risk assessment is highly technical, but the case studies, which comprise about 70 per cent of the 1120 pages of this massive volume, are immensely valuable portraits of the practice and practical applications of toxicity evaluation, exposure assessment, and risk analysis. Paustenbach's own introduction and first chapter are interesting historical surveys and contain excellent bibliographic material pertaining to almost any imaginable topic in risk analysis. *Toxic Substances and Human Risk: Principles of Data Interpretation* (R.G. Tardiff and J.V. Rodricks, Eds., Plenum Press, New York, 1987) is intended for advanced students of the topics.

Most of the volumes listed above emphasize principles and methods. Like most textbooks they are less useful as sources of up-to-date information on individual chemical substances. Several volumes are particularly valuable compilations of chemical and toxicological data on individual chemicals and commercial products. *Clinical Toxicology of Commercial Products, Fifth Edition* (R.E. Gosselin, R.P. Smith and H.C. Hodge, Editors, with the assistance of J.E. Braddock, Williams & Wilkins, Baltimore, 1984) contains information essential for the diagnosis and treatment of poisoning. Thousands of individual chemicals, drugs, pesticides, and commercial products (many with trade names) are covered. Emphasis is on acute toxicity. *Dangerous Properties of Industrial Materials*, 6th Edition (N. Irving Sax, Editor, Van Nostrand Reinhold, New York, 1984) contains not only toxicity data, but information on other types of chemical hazards (explosivity, reactivity). The volume is updated by a bimonthly journal of the same name. A multiauthor work, *Patty's Industrial Hygiene and Toxicology*, now in its 3rd edition (John Wiley and Sons, New York, 1981–2) and under the general editorship of George Clayton and Florence Clayton, contains chapters on the toxic properties of all classes of organic and inorganic chemicals. *Pesticides Studied in Man* (Williams & Wilkins, Baltimore, 1982) is a comprehensive, though now somewhat dated, review by an early master of this topic, Wayland J. Hayes. An organization called the American Conference of Governmental Industrial Hygienists (ACGIH), located in Cincinnati, Ohio, publishes annually a small volume with the long title, *Threshold Limit Values for Chemical Substances and Physical Agents and Biological Exposure Indices*. The volume lists, among other things, recommendations for limits on exposures to chemicals in the workplace; documentation for those limits is not included (it is, however, available from ACGIH). The TLVs are widely used by industrial hygienists.

An organization called Medical Economics Co., Inc., Oradell, New Jersey, publishes the *Physician's Desk Reference (PDR)*. The PDR contains information on prescription drugs, with their full labeling, as approved by FDA. Toxicity data (side effects) are included. A similar volume on OTCs (over-the-counter drugs) is also available from the same publisher.

The EPA's Office of Research and Development publishes toxicology evaluations on important industrial chemicals and pollutants, usually called 'Health Assessment' documents. Other offices of EPA issue similar reports. In the past three years another unit of the federal government, the Agency for Toxic Substances and Disease Registry (ATSDR) of the Centers for Disease Control (Atlanta, Georgia), is issuing more than 200 toxicity 'profiles' on chemicals found at hazardous waste sites; these documents also contain information on sources of the chemicals and the extent of known human exposure to them. Each ATSDR profile contains a plain language summary. Most university and large city libraries can easily obtain these EPA and ATSDR documents. All of the agencies that deal with chemicals – EPA, ATSDR, FDA, OSHA, CPSC, Department of Transportation, and their counterparts in individual states, are immensely valuable sources of information, usually obtainable at relatively small cost. All have information offices.

The World Health Organization in Geneva has since 1973 published so-called *Environmental Health Criteria* on individual chemicals and groups of related chemicals. These comprehensive reviews are available from WHO's Distribution and Sales Unit, 1211 Geneva 27, Switzerland. WHO expert committees on food additives and contaminants, pharmaceuticals, and pesticides, have also issued toxicology reviews and related information on these classes of agents. A unit of WHO, the International Agency for Research on Cancer (IARC), issues reports on carcinogens (see below, under Chapter 7).

Finally, a number of professional and trade associations publish periodicals and other information relating to chemical hazards and risks. Lists of available publications can be obtained by writing: Air Pollution Control Association, P.O. Box 2861, Pittsburgh, PA 15230; American Chemical Society, 1155 16th Street, N.W., Washington, D.C. 20036; American College of Toxicology, 9650 Rockville Pike, Bethesda, MD 20814, American Industrial Health Council, 1330 Connecticut Avenue, N.W., Washington D.C. 20036; American Petroleum Institute, 1220 L Street, N.W., Washington, D.C. 20005; American Public Health Association, 1015 15th Street, N.W., Washington, D.C. 20005; Association of State and Territorial Health Officials, 1311-A Dolly Madison Blvd., McLean, Virginia 22101; Chemical Manufacturers Association, 2501 M Street, N.W., Washington, D.C. 20037; Grocery Manufacturers of America, Inc., 1010 Wisconsin Avenue, N.W., Washington, D.C.

20007; Hazardous Waste Treatment Council, 1919 Pennsylvania Avenue, N.W., Washington, D.C. 20006; International Life Sciences Institute/Risk Science Institute; 1126 16th Street, N.W., Washington, D.C. 20036; National Governor's Association, 444 North Capitol Street, Washington, D.C. 20001; National Agricultural Chemicals Association, 1155 15th Street, N.W., Suite 900, Washington, D.C. 20005; Non-Prescription Drug Manufacturer's Association, 1150 Connecticut Avenue, N.W., Suite 1200, Washington, D.C. 20036; Pharmaceutical Manufacturers Association, 1100 15th Street, N.W., Suite 900, Washington, D.C. 20005; Public Health Foundation, 1220 L Street, N.W., Washington, D.C. 20005; Society of Toxicology, P.O. Box 90721, Washington, D.C. 20090; Society for Risk Analysis, 8000 Westpark Drive, McLean, Virginia 22101; Water Pollution Control Federation, 2626 Pennsylvania Avenue, N.W., Washington, D.C. 20037. The Chemical Industry Institute of Toxicology and the National Institute of Environmental Health Sciences (part of NIH), both located in Research Triangle Park, N.C., the Canadian Centre for Toxicology, Guelph, Ontario, Canada, and the British Industrial Biological Research Association, Carshalton, Surrey, U.K., are major research centers in toxicology. Four important advocacy groups particularly active in the area of chemical hazards are Center for Science in the Public Interest, 1875 Connecticut Avenue, N.W., Suite 300, Washington, D.C. 20009; Environmental Defense Fund, 1616 P Street, N.W., Washington, D.C. 20036; Health Research Group 2000 P Street, N.W., Suite 700, Washington, D.C. 20036; and Natural Resources Defense Council, 122 East 42nd Street, New York, New York 10168.

Prologue

Little reliable information on aflatoxin, in particular, or mold toxicants, in general, is available except in some highly technical volumes in the primary scientific literature. The chapters on 'Food Additives and Contaminants' (J.R. Hayes and T.C. Campbell) and 'Chemical Carcinogens' (G.M. Williams and J.H. Weisburger) in *Casarett and Doull's Toxicology* (op. cit.) contain substantial information on the toxicology issues. A review by D.L. Park and L. Stoloff published in the journal, *Regulatory Toxicology and Pharmacology* (Vol. 9, pages 109–30, 1989) entitled 'Aflatoxin control – how a regulatory agency managed risk from an unavoidable natural toxicant in food and feed' covers history, science, and regulatory policy. The serious student needs to locate either of two comprehensive treatises: *Mycotoxins in Human and Animal Health*, (J.V. Rodricks, C.W. Hesseltine, and M.A. Mehlman, Editors, Pathotox Publishers, Park Forest South, IL, 1977); or *Mycotoxic Fungi, Mycotoxins, Mycotoxicoses* (T.D. Wyllie and L.G. Morehouse, Editors, 3 volumes, Marcel Dekker, New York, 1977–8).

Chapter 1

For basic chemical information on commercially important products consult *The Merck Index: An Encyclopedia of Chemicals, Drugs, and Biologicals*. This one-volume and highly reliable work is in its Eleventh edition (1989), and is available from Merck & Co., Inc., Rahway, New Jersey.

Eugene Meyer's textbook *Chemistry of Hazardous Materials*, 2nd edition (Prentice Hall, Englewood Cliffs, NJ, 1989) is a primer on the chemical industry and on the chemical and physical properties of environmentally important chemicals. The National Science Foundation's report *Chemistry and the Environment*–1988 (J.W. Frost and D.M. Golden, Editors, National Science Foundation, Washington, D.C. 1988) is valuable not only for its discussion of the shape of the modern chemical industry and its contribution to environmental problems, but also for its even more extensive review of how chemical science is contributing to our understanding of those problems. Although superficial and in places somewhat alarmist in its discussion of toxic hazards and risks, L.N. Davis' *The Corporate Alchemists: Profit Takers and Problem Makers in the Chemical Industry* (William Morrow and Company, Inc., New York, 1984) provides an interesting and highly readable account of the rise and growth of the chemical industry, from dyes to explosives to drugs to pesticides. H.M. Leicester's *Historical Background of Chemistry*, first published in 1956 and available in paperback reprint from Dover Publications, Inc., New York, provides a particularly clear account of the revolution in organic chemistry of the nineteenth century. Additional historical perspective can be gained from the trustworthy Isaac Asimov in *The New Guide to Science*, 2nd Edition (Basic Books, Inc., New York, 1984).

Chapters 2 and 3

It is difficult to locate materials on exposure and dose outside of government publications and a few highly technical works. Dennis Paustenbach's book of case studies (op. cit.) has several chapters that are particularly instructive. The chapters by S.M. Brett *et alia* 'Assessment of the Public Health Risks Associated with the Proposed Excavation of a Hazardous Waste Site'; P.K. LaGoy *et alia*, 'The Endangerment Assessment of the Smuggler Mountain Site, Pitkin County, Colorado: A Case Study'; R.G. Gammage and C.C. Travis, 'Formaldehyde Exposure and Risk in Mobile Homes'; and L. Tollefson, 'Methylmercury in Fish: Assessment of Risk for U.S. Consumers', all contain discussions of the types of data necessary to evaluate exposure and intakes of environmental chemicals, from several different environmental media. The EPA's Office of Remedial Response has issued the *Superfund Exposure Assessment Manual* (Washington, D.C., 1988, EPA/540/1-88/001). This document not only provides guidance on the conduct of human exposure assessments for substances migrating from uncontrolled hazardous waste sites, but provides substantial background discussion on the movement of chemicals in the environment, the rates of human contact with various environmental media, and on the integration of information to produce dose estimates. The World Health Organization's *Guidelines for Predicting Dietary Intake of Pesticide Residues* (Geneva, 1989) can be consulted for information on how this important topic is generally handled.

Two chapters in *Casarett and Doull's Toxicology* (op. cit.), 'Distribution, Excretion, and Absorption of Toxicants' (C.D. Klaassen) and 'Biotransformation of Toxicants' (I.G. Sipes and A.J. Gandolfi) get to the details of how chemicals move in, around, and out of the body, and of how they are metabolized when in it. Unfortunately, most of this material requires fairly advanced understanding of biology and organic chemistry.

Chapter 4

There appears to be no recent, general history of the toxicological sciences available. The 1959 volume by K.P. Dubois and E.M.K. Geiling, *Textbook of Toxicology* (Oxford University Press) contains a fair amount of historical information. Alice Hamilton's *Exploring the Dangerous Trades* (Little Brown Co., Boston, 1943) was reprinted in 1985 by Northeastern University Press, Boston. *Silent Spring* by Rachel Carson was published in 1962 by Houghton Mifflin Co., Boston, and numerous paperback editions have been issued. Hamilton's and Carson's books are both historically important because they focussed public attention on toxicological issues. An early piece on carcinogens, 'Cancer and Environment', by Groff Conklin, appeared in *Scientific American* in January , 1949; it has been reprinted in a volume called *Cancer Biology* in the 'Readings from Scientific American' series (W.H. Freeman and Company, New York). Wilhelm Hueper published several major works on chemical carcinogens in the 1940–65 period; consult his *Occupational Tumors and Allied Diseases* (C.C. Thomas, Springfield, IL, 1942) for much material of historical interest. E.B. Lewis' 1957 publication in the journal *Science*, entitled 'Leukemia and Ionizing Radiation' (Vol., 125, No. 3255, May, 1957) was an important, and early contribution to the linear, no-threshold hypothesis for carcinogens that eventually spilled over from radiation to chemicals.

Chapters 5 and 6

Sources of toxicity data on individual chemicals were cited above. In addition to those the multiauthor volume, edited by A. Wallace Hayes. *Principles and*

Methods of Toxicology (Raven Press, New York, 1984) is an excellent source of information on the design, conduct, and interpretation of animal tests for toxicity. A series of relatively inexpensive books from the National Academy of Sciences/National Research Council entitled *Drinking Water and Health* (nine volumes, 1977–present) is useful both for its discussions of toxicological principles and for its review and evaluation of individual chemicals; it also provides an important perspective on the problem of pollutants in drinking water. *Toxic Chemicals, Health, and the Environment* edited by L.B. Lave and A.C. Upton (The Johns Hopkins University Press, Baltimore, 1987) covers a wide range of toxicology issues in a number of well-written and fairly accessible chapters. *Food Safety* (H. Roberts, Editor, John Wiley and Sons, New York, 1981) offers in relatively concise form a view of the risks of all the classes of dietary constituents, additives and contaminants. More comprehensive and technical, and containing extensive discussions of toxicological principles, is *Food Toxicology* by Jose M. Concon (Marcel Dekker, New York, 1988). The special topic of effects of chemicals on reproduction and on the developing organism is well-covered in *Reproductive Hazards of Industrial Chemicals, and Evaluation of Animal and Human Data* (S.M. Barlow and F.M. Sullivan, Academic Press, New York, 1982). See also the various chapters in *Casarett and Doull's Toxicology* (op. cit.) on target organ toxicity. R.A. Goyer's review of 'Environmentally Related Diseases of the Urinary Tract' is a fine example of a toxicology review written for clinicians (*Medical Clinics of North America*, Vol. 74, No. 2, March, 1990, pp 377–89). An interesting review of newly emerging issues in toxicology is provided by G.E. Simon, W.J. Katon, and P.J. Sparks in 'Allergic to Life: Psychological Factors in Environmental Illness', *American Journal of Psychiatry*, Vol. 147, No. 7, July, 1990, pp 901–6). *Chemical Hazards of the Workplace*, 2nd Edition, N.H. Proctor, J.P. Hughes and M. Fischman, Lippincott, Philadelphia, PA, 1988) is a reasonably thorough and readable discussion of that topic. For those with some training in statistics see the classic work by C.L. Bliss, *The Statistics of Bioassays* (Academic Press, New York, 1952). Another classic and comprehensive treatment of statistics in biology is offered by P. Armitage in *Statistical Methods in Medical Research* (Blackwell Scientific Publications, Oxford, 1971, and reprinted several times since).

Chapters 7 and 8

The literature on cancer and chemical carcinogens is massive. A good place to begin is 'The Cancer Problem' by John Cairns, published in *Scientific American* in November, 1975, and reprinted in *Cancer Biology* (op. cit.). Although it concerns radiation, Arthur Upton's 1982 article in the same collection 'The Biological Effects of Low-Level Ionizing Radiation' is an especially clear introduction to mechanisms of carcinogenesis and risk assessment. The entire volume of readings is recommended for the serious student. A fine and nontechnical discussion of the cancer problem is provided in a 1980 paperback publication by the National Cancer Institute, National Institute of Health, entitled *Science and Cancer*. Its author is the noted cancer biologist Michael B. Shimkin. Henry Pitot's *Fundamentals of Oncology*, 3rd Edition Revised and Expanded (Marcel Dekker, Inc., New York, 1986) is superb, but is strictly for the advanced student.

The next stopping point might be *Assessment of Technologies for Determining the Cancer Risks from the Environment*, a comprehensive and clearly-written report issued in 1981 by the Office of Technology Assessment (OTA), an advisory arm of the U.S. Congress. It is available from the U.S. Government Printing Office, Washington, D.C. This is an immensely useful guide not only to scientific fundamentals, but also to regulation. The basic expidemiology of cancer and Doll and Peto's estimates of the causes of avoidable risks of cancer in the United States, and other such estimates, are extensively reviewed. The paper by D.L. Davis and co-authors entitled 'International trends in cancer mortality in France, West Germany, Italy, Japan, England and Wales, and the USA' (*The Lancet*, Vol. 336, pages 474–81, 1990) provides additional and more recent analyses of trends in cancer rates.

The New York Academy of Sciences, New York City, can provide *Living in a Chemical World: Occupational and Environmental Significance of Industrial Carcinogens* (Cesare Maltoni and Irving J. Selikoff, Editors, Volume 534 of the Annals of the New York Academy of Sciences, 1988). This multiauthor volume compiles papers given at an international conference held by the Collegium Ramazzini in Bologna, Italy, in October, 1985. It is valuable because it contains several papers on the work of important organizations such as the IARC, the International Agency for Research on Cancer (see the papers by Lorenzo Tomatis, page 31 and William Nicholson, page 44), IPCS, the International Program on Chemical Safety (E. Smith's paper on page 39), and NTP, the National Toxicology Program of the U.S. Government (paper by J.E. Huff *et alia*, page 1). These four papers provide additional references to important IARC, IPCS and NTP publications. David Rall's paper in the same volume, 'Laboratory animal toxicity and carcinogenesis testing: underlying concepts, advantages, and constraints' on page 78 is a clear and concise discussion of cancer bioassays.

Turn again to *Casarett and Doull's Toxicology* (op. cit.) for the excellent chapter on 'Chemical Carcinogens' by Gary M. Williams and John H. Weisburger. Two highly technical, multiauthor volumes worth consulting are *Handbook of Carcinogen Testing* (H.A. Milman and E.K. Weisburger, Editors, Rogers Publications, Park Ridge, N.J., 1985) and *Chemical Carcinogens*, Second Edition, Revised and expanded (Charles E. Searle, Editor, ACS Monograph 182, American Chemical Society, Washington, D.C., 1984, 2 volumes). The first of these two works is principally devoted to tests for identifying carcinogens while the second is concerned with the classes of chemical carcinogens. Both are packed with information and references. The former work also contains a fine review entitled 'Role of epidemiology in identifying chemical carcinogens' by C.S. Muir and John Higginson, pages 28–54. The 1981 Office of Technology Assessment report (op. cit.) also provides a good introduction to the methods of epidemiology applicable to identifying carcinogens. Alvin R. Feinstein's review 'Scientific standards in epidemiologic studies of the menace of daily life' (*Science*, Vol. 242, 2 December, 1988, pages 1257–63) is a discussion of the pitfalls of the epidemiologic method.

Discussions of mechanisms of carcinogenesis can be found in many of the above references, see particularly the Williams and Weisburger chapter in *Casarett and Doull's Toxicology*. A concise and clear presentation of this topic is also provided by David B. Clayson in a paper entitled 'Can a mechanistic rationale be provided for non-genotoxic carcinogens identified in rodent bioassays?' (*Mutation Research*, Vol. 221, 1989, pp 53–67). This paper was prepared for the International Commission for Protection Against Environmental Mutagens and Carcinogens, and sets forth some of the reasons for believing that not all chemical carcinogens act through 'no-threshold' mechanisms. Samuel M. Cohen and Leon B. Ellwein's article in *Science* (Vol., 249, 31 August, 1990, pp 1007–11) entitled 'Cell proliferation in carcinogenesis', pursues the same objective.

Justification for the use of the MTD in cancer studies is discussed by David Rall (op. cit.) and in the OTA report (op. cit.). Bruce Ames and Lois Swirsky Gold present a contrary review in 'Too many rodent carcinogens: mitogenesis increases mutagenesis' (*Science*, Volume 249, 31 August, 1990, pp 970–1).

Two historically-important publications on mechanisms of carcinogenesis are Berenblum and Shubik's 1947 paper 'A new, quantitative approach to the study of the stages of chemical carcinogenesis in the mouse's skin' (*British Journal of Cancer*, Volume 1, pages 303–91) and Armitage and Doll's 'The age distribution of cancer and a multi-stage theory of carcinogenesis,' also published in the *British Journal of Cancer* (Volume 8, pages 1–12, 1954).

The journalist Edith Efron published *The Apocalyptics: Cancer and the Big Lie* (Simon and Schuster, New York, 1984) to 'expose' how politically-motivated scientists distort evidence and create unjustifiable alarms about carcinogens in the environment. The tone of Efron's book is one of anger and its message conspiratorial. Although it is highly repetitive, it is packed with information and expresses a view held by many scientists.

One of the 'alarmist' scientists singled-out by Efron is Samuel Epstein, whose book, *The Politics of Cancer* (revised edition, Anchor Press/Doubleday, Garden City, New York, 1979) attempts to expose industrial and governmental negligence regarding cancer risks in the environment. The Efron–Epstein volumes combine to create an interesting but unfortunately, highly confusing picture of the underlying scientific issues.

Chapter 9

Most of the general reference works cited above contain discussions of the dose–response issue. See especially 'Principles of Toxicology' by Curtis D. Klaassen, Chapter 2 of *Casarett and Doull's Toxicology* (op. cit.) and Rolf Hartung's chapter, 'Dose–Response Relationships' in *Toxic Substances and Human Risk* (op. cit.). The concepts of NOELS and safety factors for threshold effects are discussed at length in this same volume by J. Rodricks and R.G. Tardiff in the chapter 'Comprehensive Risk Assessment'. Frank Lu's *Basic Toxicology* (op. cit.) also deals with this topic. Extrapolation of carcinogenicity dose–response data from high-to-low doses is well-presented in the 1981 OTA report (op cit.), Chapter 5. The scientific bases for these forms of dose–response extrapolation are set forth in a report from the President's Office of Science and Technology Policy, and published in the *Federal Register*, in 1985, under the title 'Chemical carcinogens: A review of the science and its associated principals' (Volume 50, pages 10372–442). Scientists from all the major U.S. federal regulatory agencies contributed to the report. The Cohen and Ellwein article, 'Cell proliferation in carcinogenesis' *(Science* (op. cit.))* provides a discussion of the scientific basis for alternative methods of low dose extrapolation for certain classes of carcinogens.

Basic problems of animal-to-human extrapolation are in the OTA's report (op. cit.) and comprehensively treated in Edward Calabrese, *Principles of Animal Extrapolation* (John Wiley and Sons; New York, 1983). The particular problem of extrapolation of teratology data from animals to humans is concisely discussed by V. Frankos in his paper 'FDA perspectives in the use of teratology data for human risk assessment' (*Fundamental and Applied Toxicology*, Vol. 5, 1985, pp 615–25).

Chapter 10

Dennis J. Paustenbach's 'A Survey of Health Risk Assessment', Chapter 1 of *The Risk Assessment of Environment Hazards* (op. cit.) is a good place to start because it provides a reasonably current, bird's-eye view of the whole risk assessment scene. The 1983 report of the National Research Council/National Academy of Sciences, *Risk Assessment in the Federal Government: Managing the Process* (National Academy Press, Washington, D.C.) is essential reading for anyone attempting to grasp the fundamental science and policy issues in risk assessment.

The practice of risk assessment is well exhibited in many of the case studies collected in the Paustenbach volume. See particularly Linda Tollefson's 'Methylmercury in Fish: Assessment of Risks of U.S. Consumers' (Chapter 25); E. Marshall Johnson's 'A Case Study of Developmental Toxicity Risk Estimation Based on Animal Data: The Drug Bendectin' (Chapter 21); and

'Assessment of a Waste Site Contaminated with Chromium' (Chapter 10, by R.J. Golden and N.J. Karch).

Carcinogen Risk Assessment, edited by Curtis S. Travis (Plenum Press, New York, 1988) and *Toxicological Risk Assessment*, 2 volumes (D.B. Clayson, D. Krewski, and I. Munro, CRC Press, Boca Raton, Florida, 1985) are two comprehensive works emphasizing methodological issues. Both are fairly technical.

The United States federal government's view can be located in the Office of Science and Technology Policy review of 1985 (op. cit.) and the EPA's 'Guidelines for carcinogen risk assessment' published in the *Federal Register*, Vol. 51, pages 33992–4003, 1985. The State of California has also published such guidelines: *Carcinogen Identification Policy: A Statement for Estimating Cancer Risks from Exposure to Carcinogens*: (Health and Welfare Agency, Sacramento, 1982).

Chapter 11

A superb summary of the U.S. regulatory structure and of the laws that underpin it is provided by Richard Merrill in a chapter entitled 'Regulatory Toxicology', contained in *Casarett and Doull's Toxicology* (op. cit.). The National Research Council/National Academy of Sciences report *Risk Assessment in the Federal Government* (op. cit.) and the OTA's *Assessment of Technologies for Determining Cancer Risk from the Environment* (op cit.) are valuable sources of information and analysis regarding fundamentals of risk management. A brief 1988 report from the Conservation Foundation, Washington, D.C. *Risk Management and Risk Control*, is a good primer.

Management of Assessed Risks from Carcinogens (W.J. Nicholson, Editor, Annals of the New York Academy of Sciences, Vol. 363, April 30 1981) is a broad, multiauthor treatment. My own paper in this volume, 'Regulation of carcinogens in food' provides some historical perspective on food safety and also deals with the problem of naturally-occurring carcinogens.

A relatively early, and highly influential work, *Of Acceptable Risk: Science and The Determination of Safety*, by William Lowrance (William Kaufman, Los Altos, CA, 1979) sets forth the basic issues in the determination of safety. Also widely-cited is Peter Barton Hutt's 'Legal considerations in risk assessment under federal regulatory statutes' in *Assessment and Management of Chemical Risks*, (J.V. Rodricks and R.G. Tardiff, Editors, ACS Symposium Series 239, American Chemical Society, Washington, D.C. 1984). Fred Hoerger offers an excellent perspective on the role of risk assessment in corporate decision-making in this same volume (Chapter 10).

See also *Acceptable Risk* (Branch Fischoff, Sarah Lichtenstein, Paul Slovic, Stephen L. Denby, and Ralph L. Kenney, Cambridge University Press, New York, 1989) for information on and insights into the problems of risk perception.

Theodore S. Glickman and Michael Gough of Resources for the Future (RFF, 1990), Washington, D.C., have compiled *Readings in Risk*, a collection of previously published papers on all aspects of risk and its assessment and management, chosen because of their influence among practitioners of risk assessment and policy makers. The volume is available from RFF. Most of the papers are well worth study. Chauncey Starr's 'Social Benefit Versus Technological Risk', first published in 1969, is a thoughtful discussion of risk-benefit analysis. Richard Wilson's review 'Analyzing the Daily Risks of Life' has been heavily cited by policy makers. M. Granger Morgan contributes two superb papers on understanding and managing the risks of technology. Problems of risk communication are thoughtfully examined in papers by Peter M. Sandman and several others. Also consult a fascinating piece of policy analysis by Aaron Wildavsky called 'No Risk is the Highest Risk of All'. William Ruckelshaus, former Administrator (twice) of EPA is represented by a fine article on the dilemmas facing the regulator, 'Risk, Science, and Democracy'. These and many other articles make the Glickman–Gough volume essential reading. Michael Gough has himself contributed significantly to the debate over managing risks from environmental carcinogens: see his article 'How much cancer can EPA regulate away' in *Risk Analysis*, Volume 10, pages 1–7, 1990. *In Search of Safety* (John D. Graham, Laura C. Green, and Marc J. Roberts, Harvard University Press, Cambridge, MA, 1988) is an interesting analysis of the interaction of science and regulatory policy.

Significant cross-national differences in dealing with questions of risk are discussed by Shelia Jasanoff in 'American exceptionalism and the political acknowledgement of risk' in the fall, 1990, issue of *Daedalus* (American Academy of Arts and Sciences, Cambridge, MA). See also Mary Douglas' 'Risk as a forensic resource' and Stephen Klaidman's 'How well the media report health risks' in the same issue. The whole issue is devoted to the subject of risk as a social and political phenomenon.

Additional citations

The quoted materials on pages 10 and 169 are from *Living in a Chemical World: Occupational and Environmental Significance of Industrial Carcinogens* (op. cit.). The Shimkin quote on pages 109–10 is from *Science and Cancer* (op. cit.). *Silent Spring* (op. cit.) is the source for the three brief Rachel Carson quotes (pages 43 and 114). Williams and Weisburger's definition of a carcinogen is from *Casarett's and Doull's Toxicology: The Basic Science of Poisons*, 3rd Ed. (op. cit.). My own remarks on page 229 are from *Management of Assessed Risks from Carcinogens* (op. cit.).

Note added in proof: *Casarett and Doull's Toxicology: The Basic Science of Poisons* (M.O. Amdur, J. Doull, and C.D. Klassen, Eds.) has recently been issued in a Fourth Edition by Pergamon Press, New York. (1991).

Index